Illuminating the Hidden Planet

THE FUTURE OF SEAFLOOR OBSERVATORY SCIENCE

COMMITTEE ON SEAFLOOR OBSERVATORIES: CHALLENGES AND OPPORTUNITIES

OCEAN STUDIES BOARD
COMMISSION ON GEOSCIENCES, ENVIRONMENT, AND RESOURCES
NATIONAL RESEARCH COUNCIL

NATIONAL ACADEMY PRESS
WASHINGTON, D.C.

NATIONAL ACADEMY PRESS • 2101 Constitution Avenue, N.W. • Washington, DC 20418

NOTICE: The project that is the subject of this report was approved by the Governing Board of the National Research Council, whose members are drawn from the councils of the National Academy of Sciences, the National Academy of Engineering, and the Institute of Medicine. The members of the committee responsible for the report were chosen for their special competencies and with regard for appropriate balance.

This report and the committee were supported by a grant from the National Science Foundation. The views expressed herein are those of the authors and do not necessarily reflect the views of the sponsors.

International Standard Book Number 0-309-07076-7

Additional copies of this report are available from the National Academy Press, 2101 Constitution Ave., N.W., Lockbox 285, Washington, D.C. 20055; 1-800-624-6242 or 202-334-3313 (in the Washington metropolitan area); Internat: http://www.nap.edu.

Printed in the United States of America.

THE NATIONAL ACADEMIES

National Academy of Sciences
National Academy of Engineering
Institute of Medicine
National Research Council

The **National Academy of Sciences** is a private, nonprofit, self-perpetuating society of distinguished scholars engaged in scientific and engineering research, dedicated to the furtherance of science and technology and to their use for the general welfare. Upon the authority of the charter granted to it by the Congress in 1863, the Academy has a mandate that requires it to advise the federal government on scientific and technical matters. Dr. Bruce M. Alberts is president of the National Academy of Sciences.

The **National Academy of Engineering** was established in 1964, under the charter of the National Academy of Sciences, as a parallel organization of outstanding engineers. It is autonomous in its administration and in the selection of its members, sharing with the National Academy of Sciences the responsibility for advising the federal government. The National Academy of Engineering also sponsors engineering programs aimed at meeting national needs, encourages education and research, and recognizes the superior achievements of engineers. Dr. William A. Wulf is president of the National Academy of Engineering.

The **Institute of Medicine** was established in 1970 by the National Academy of Sciences to secure the services of eminent members of appropriate professions in the examination of policy matters pertaining to the health of the public. The Institute acts under the responsibility given to the National Academy of Sciences by its congressional charter to be an adviser to the federal government and, upon its own initiative, to identify issues of medical care, research, and education. Dr. Kenneth I. Shine is president of the Institute of Medicine.

The **National Research Council** was organized by the National Academy of Sciences in 1916 to associate the broad community of science and technology with the Academy's purposes of furthering knowledge and advising the federal government. Functioning in accordance with general policies determined by the Academy, the Council has become the principal operating agency of both the National Academy of Sciences and the National Academy of Engineering in providing services to the government, the public, and the scientific and engineering communities. The Council is administered jointly by both Academies and the Institute of Medicine. Dr. Bruce M. Alberts and Dr. William A. Wulf are chairman and vice chairman, respectively, of the National Research Council.

STEERING COMMITTEE

WILLIAM B. F. RYAN, *Chair*, Lamont-Doherty Earth Observatory, Palisades, New York
ROBERT DETRICK, *Vice-Chair*, Woods Hole Oceanographic Institution, Massachusetts
KEIR BECKER, University of Miami, Florida
JAMES BELLINGHAM, Monterey Bay Aquarium Research Institute, Moss Landing, California
ROGER LUKAS, University of Hawaii, Manoa
JOHN LUPTON, NOAA-Pacific Marine Environmental Laboratory, Newport, Oregon
LAUREN MULLINEAUX, Woods Hole Oceanographic Institution, Massachusetts
JACK SIPRESS, Sipress Associates, Holmdel, New Jersey

Staff

ALEXANDRA ISERN, Study Director
SHARI MAGUIRE, Research Assistant
MEGAN KELLY, Project Assistant

COMMISSION ON GEOSCIENCES, ENVIRONMENT, AND RESOURCES

Staff

Foreword

Much of what we know about the ocean is the result of ship-based expeditionary science with a tradition dating back to the late 19th century. But, it has become apparent that to answer many important questions in ocean and earth science, a coordinated research effort of long-term investigations of ocean processes is required. Seafloor observatories can provide some of the necessary data sets, while also providing a full-time presence on the seafloor in a variety of environments that have the capability to capture public imagination. Much of the research that is proposed for ocean observatories is interdisciplinary in nature and has the potential to greatly advance oceanographic science at a time when understanding our ocean is of increasing importance to society.

This report, *Illuminating the Hidden Planet: The Future of Seafloor Observatory Science*, continues the efforts of the Ocean Studies Board (OSB) in encouraging interdisciplinary research through many of its reports (NRC, 1998a; NRC, 1999; NRC, 2000). The OSB recognizes that interdisciplinary research plays an important role in ensuring continued vitality of the ocean science community.

The significant scientific advances resulting from existing seafloor observatory initiatives have shown the great intellectual potential of observatory science. Recognition of this has encouraged the academic community to begin planning for a major seafloor observatory program. This report encourages the establishment of a seafloor observatory network and provides recommendations to help guide this effort.

The seafloor observatory initiative is coming together at a very propitious time, when quantitative interdisciplinary ocean science has become possible on a broad basis. This is an exciting moment for the ocean community.

Kenneth Brink
Chair, Ocean Studies Board

Preface

The Committee on Seafloor Observatories was charged with examining the scientific merit of, technical requirements for, and overall feasibility of establishing a system of seafloor observatories. For the purpose of this report, seafloor observatories are defined as unmanned, fixed systems of instruments, sensors, and command modules connected either acoustically or via a seafloor junction box to a surface buoy or a fiber optic cable to land. These observatories will have power and communication capabilities and will provide support for spatially distributed sensing systems and mobile platforms. Sensors and instruments that are utilized at seafloor observatories will have the potential to collect data from above the air-sea interface to below the seafloor and will provide support for in situ manipulative experiments.

An important component of this study was to gauge the level of support for observatory science within the ocean and earth science communities. To this end, the "Symposium on Seafloor Observatories" was designed as an opportunity for comprehensive discussions on the scientific potential and technical needs associated with the establishment of a network of seafloor observatories. This meeting, held in Islamorada, Florida, from January 10-12, 2000, brought together prominent researchers and representatives from all fields of ocean and earth science, and from disciplines as diverse as engineering and planetary exploration. The diverse backgrounds of the attendees meant that there was extensive interdisciplinary interaction. Attendees at the symposium represented overseas institutions, government agencies, private industry, and academia. The committee used the high number of attendees and the very positive response to the symposium by the participants as a measure of the strong community support for a seafloor observatory program.

To encourage active discussions on the scientific potential of seafloor observatories, the Symposium began with keynote lectures on the role of sustained time-series observations for advancing earth and ocean science, presented by a number of distinguished speakers. The topics for these presentations were interdisciplinary in nature and representative of a wide range of research and technical areas. Keynote presentations were followed by a series of breakout groups. The first two breakout groups were divided along interdisciplinary scientific themes. The charges to these groups were designed to determine the potential for seafloor observatories to lead to significant scientific advances in ocean and earth science, and to discuss observatory architectures best suited for investigating these scientific problems. The third breakout group was divided according to the three main observatory types that had been proposed: relocatable, long-term, and global/basin-scale observatories. The charge for this breakout group was to identify the technical requirements needed to establish a network of seafloor observatories. An outcome of these technical discussions was a reformulation of the observatory types originally proposed as it was felt that the main observatory architectures consisted of cable- and mooring-based systems with relocatable systems being a subset of moored-buoy observatories. This new classification is followed in this report. Posters displayed at the symposium demonstrated the extensive multidisciplinary observatory research currently underway and also showed exciting new scientific areas that can be addressed with long time-series data sets collected with seafloor observatories.

Discussions during general sessions, within breakout groups, and at the poster sessions helped provide a clear picture of both the scientific rationale and technical requirements needed to establish an observatory network. An exciting outcome of symposium discussions was recognition of the many interdisciplinary synergies in both measurement type and location. Discussions at the symposium also helped define engineering capabilities required for the establishment of a seafloor observatory network, including identification of specific fields where technological advances are needed. Throughout symposium discussions, the potential scientific and resource trade-offs involved in pursuing a large program of seafloor observatories were recognized. During subsequent committee discussions, a prudent stepwise approach to their establishment was recommended.

The purpose of this report is to discuss the scientific rationale and technical feasibility of establishing a seafloor observatory network. The recommendations discussed here will provide the National Science Foundation, and other agencies involved in observatory science, with advice on whether and how to proceed with the establishment of an ocean observatory network. In addition, this document is intended to supply the scientific community with information on observatory science and to initiate further discussion on the prominent scientific problems that can only be addressed with long time-series measurements.

Chapter 1 of this report provides an overview of the rationale for establishing a network of seafloor observatories, a description of the types of observatories discussed in the report, and a description of major proposed and active observatory programs. Chapter 2 discusses the potential of seafloor observatories to address important research areas in ocean and earth science. Chapter 3 describes available technology and the technical development needed to establish the proposed seafloor observatory architectures while Chapter 4 discusses other technical requirements that will be needed. Chapter 5 presents additional issues of significance for establishing a seafloor observatory program, such as project management, data management, and public outreach. Chapter 6 lists the Committee findings and recommendations.

William B. F. Ryan
Chair, Committee on Seafloor Observatories

Acknowledgments

This report and the well-attended and productive Symposium on Seafloor Observatories were greatly enhanced by the symposium participants. In particular, the steering committee would like to acknowledge the efforts of those who gave keynote presentations at the symposium: Carl Wunsch, Stace Beaulieu, Meg Tivey, Anna-Louise Reysenbach, Peter Jumars, Adam Dziewonski, Patrick Trischitta, Dan Frye, Alan Chave, Ken Johnson, Dana Yoerger, and Herb Kroehl. These talks helped set the stage for fruitful discussions in the breakout sessions that followed. The steering committee is also grateful to those who led breakout group discussions: Robert Weller, Miriam Kastner, Robin Bell, John Orcutt, Keith Raybould, Chris Fox, Ken Smith, Tommy Dickey, Fred Duennebier, Marv Lilley, Barbara Romanowicz, and Doug Luther. John Delaney is gratefully acknowledged for his enjoyable and informative presentation after the symposium dinner. The efforts of Megan Kelly of the Ocean Studies Board are acknowledged for making the Symposium a well-organized and enjoyable event, and Shari Maguire for ensuring the report process ran smoothly.

This report has been reviewed in draft form by individuals chosen for their diverse perspectives and technical expertise, in accordance with procedures approved by the NRC's Report Review Committee. The purpose of this independent review is to provide candid and critical comments that will assist the NRC in making the published report as sound as possible, and to ensure that the report meets NRC standards for objectivity, evidence, and responsiveness to the study charge. The review comments and draft manuscript remain confidential to protect the integrity of the deliberative process. The committee wishes to thank the following individuals for their participation in the review of this report: Alice Alldredge (University of California, Santa Barbara), Daniel Frye (Woods Hole Oceanographic Institution), Nelson Hogg (Woods Hole Oceanographic Institution), Gary Klinkhammer (Oregon State University), Herbert Kroehl (National Oceanic and Atmospheric Administration), Marcia McNutt (Monterey Bay Aquarium Research Institute), Brad Mooney (J. Brad Mooney Associates, Ltd.), Barry Raleigh (University of Hawaii, Manoa), Fred Noel Spiess (University of California, San Diego), James Wenzel (Marine Development Associates, Inc.), Carl Wunsch (Massachusetts Institute of Technology), and Dana Yoerger (Woods Hole Oceanographic Institution). While the individuals listed above have provided constructive comments and suggestions, it must be emphasized that responsibility for the final content of this report rests entirely with the authoring committee and the NRC.

Contents

Executive Summary

Earth's oceans are essential to society as a source of food and minerals, a place of recreation, an economic means of transporting goods, and a keystone of our national security. Despite our reliance on the ocean and its resources, it remains a frontier for scientific exploration and discovery. Scientists have been using ships to explore the ocean with great success over the past 50 years and this mode of expeditionary science has led to remarkable increases in oceanographic knowledge. A ship-based expeditionary approach, however, is poorly suited for investigating changes in the ocean environment over extended intervals of time. To advance oceanographic science further, long time-series measurements of critical ocean parameters, such as those collected using seafloor observatories, are needed (NRC, 1998a).

For the purpose of this report the term "seafloor observatories" is used to describe an unmanned system of instruments, sensors, and command modules connected either acoustically or via a seafloor junction box to a surface buoy or a fiber optic cable to land. These observatories will have power and communication capabilities and will provide support for spatially distributed sensing systems and mobile platforms. Instruments and sensors will have the potential to make measurements from above the air-sea interface to below the seafloor and will provide support for in situ manipulative experiments.

In the fall of 1999, the National Science Foundation (NSF) asked the National Research Council (NRC) to investigate the scientific merit, technical requirements, and overall feasibility of establishing the infrastructure needed for a network of unmanned seafloor observatories. In addition, NSF asked the NRC to (1) assess the extent to which seafloor observatories will address future requirements for conducting multidisciplinary research in the oceans and (2) gauge the level of support for observatory science within the ocean sciences and the broader scientific community. The NRC Committee on Seafloor Observatories was appointed to carry out this charge. The Committee's findings and recommendations are based on the knowledge and experience of Committee members, consideration of various reports and workshop documents, and input from the "Symposium on Seafloor Observatories" held in January 2000.

The Committee concludes that seafloor observatories present a promising and in some cases essential new approach for advancing basic research in the oceans, and encourages NSF to move ahead with plans for a seafloor observatory program. Furthermore, based on written and verbal responses from the symposium, the Committee views the ocean and earth science community as being enthusiastic and supportive of seafloor observatory research. It should be noted that the strong multidisciplinary support for observatories is based on an observatory concept that encompasses a wide spectrum of facilities and substantial flexibility in their geographic positioning. An observatory system restricted to a single facility type or a discipline-specific geographic focus would not garner the same broad enthusiasm. The Committee also cautions that, although seafloor observatories provide a significant scientific opportunity, there are risks involved in this endeavor. Both potential benefits and risks are outlined in this report.

The scientific benefit of seafloor observatory investigations has been recognized for many years and, as such, numerous independent national and international observatory efforts have been proposed or are underway. Many of these efforts are described here. Although the seafloor observatory program discussed in this report will be primarily research driven, the data collected by the proposed observatories will provide an important contribution to operational observing systems, such as the international Global Ocean Observing System (GOOS).

SCIENTIFIC MERIT OF SEAFLOOR OBSERVATORIES

Seafloor observatories could offer earth and ocean scientists unique new opportunities to study multiple, interrelated processes over timescales ranging from seconds to decades; to conduct comparative studies of regional processes and spatial characteristics; and to map whole-earth and basin-scale structures. The scientific problems driving the need for seafloor observatories are broad in scope, spanning nearly every major area of marine science. Many of these have been previously identified in the NSF long-range plan "Futures" reports (Baker and McNutt, 1996; Jumars and Hay, 1999; Mayer and Druffel, 1999; Royer and Young, 1999). Some of the most compelling of these scientific problems are discussed below.

ROLE OF THE OCEAN IN CLIMATE

The ocean influences climate through processes that vary over regional (and smaller) scales, such as equatorial upwelling, western boundary circulation, subtropical subduction, and deep convection in high latitudes. These ocean circulation processes not only participate directly in climate variability through their influence on sea surface temperature but also affect the coupled ocean-atmosphere system through their influence on ocean biogeochemistry (e.g., the carbon cycle). To understand and potentially predict climate variations on longer timescales requires time-series measurements that resolve rapid processes, such as internal waves. A seafloor observatory network that provides globally distributed, fixed location time-series measurements of relevant surface and water-column properties would be an important contribution to a systematic approach toward climate research and prediction.

FLUIDS AND LIFE IN THE OCEAN CRUST

Although ocean chemistry is greatly influenced by the movement of fluids through oceanic crust, the processes controlling this flow are poorly understood. Recent studies on the nature of the subsurface biosphere have indicated that the crust contains a population of dormant microbes that are periodically driven into a population explosion by input of heat and volatiles into the crust during magma emplacement events. Time-series measurements of fluid movement near active ridgecrest vent fields, on ridge flanks, and at convergent margins will be critical for directly observing the changes in heat, chemical fluxes, and biological diversity produced by magmatic or tectonic events, and will form the basis for understanding how these changes influence global biogeochemical cycles.

DYNAMICS OF OCEANIC LITHOSPHERE AND IMAGING EARTH'S INTERIOR

Many of Earth's dynamic tectonic systems will be difficult to understand fully without a continuous observational presence provided by the establishment of seafloor observatories. These include the complex magmatic and tectonic systems operating at ridge crests and submarine volcanoes; the genesis of destructive earthquakes and tsunamis at subduction zone megathrusts and their relationships to large-scale plate motions, strain accumulation, fault evolution, and subsurface fluid flow; the geodynamics of Earth's interior and the origin of Earth's magnetic field; and the motion and internal deformation of lithospheric plates. Seafloor observatories also have the potential to play a key role in the global assessment and monitoring of geological hazards, as many of

Earth's most seismogenic areas and most active volcanoes occur along continental margins.

COASTAL OCEAN PERTURBATION AND PROCESSES

An important factor limiting coastal ocean research is the inability to quantify vertical and horizontal transport of water, elements, and energy through the coastal ocean system. Long time-series measurements of critical parameters, such as temperature, salinity, nutrients, and trace elements, will help provide the data needed to remedy this deficiency. Furthermore, anthropogenic influences are strongly felt in the coastal ocean through such effects as excess nutrient and contaminant inputs. It is difficult to assess the full impact of these inputs without being able to quantify the fates and transports of materials through the coastal zone. These transports are critical for the understanding of such factors as how the coastal ocean influences biogeochemical cycles and how turbulence in the coastal zone influences primary productivity.

TURBULENT MIXING AND BIOPHYSICAL INTERACTIONS

Successful modeling of the distribution of organisms and chemical compounds in the ocean depends directly on the predictive quality of circulation models, which are, in turn, limited by our ability to model turbulent mixing in the ocean. Because turbulent motions result from highly nonlinear dynamics acting across a range of time and space scales, from the dissipation scale of a few millimeters to mesoscale eddies with diameters of approximately 100 km, progress in understanding these motions is difficult. Continued advances will depend on the ability to systematically collect long-term measurements resolving small vertical and horizontal scales throughout the range of turbulent regimes that are controlled by extremes of mechanical forcing, buoyancy forcing, and topography.

ECOSYSTEM DYNAMICS AND BIODIVERSITY

The biological, ecological, and biogeochemical questions likely to benefit most from sustained ocean time-series observations are those involving time-dependent processes or episodically triggered events. Important time-dependent problems include population dynamics of predators and their prey; changes in fish stocks over time; and the effects of daily, seasonal, and life-cycle migrations of populations on biogeochemical processes. In addition to the observation of natural events and perturbations, it is anticipated that the ability to conduct active experiments, such as controlled releases of chemicals or tracers into the water column or manipulations of seafloor communities, will be a beneficial outcome of observatory science. Furthermore, acquisition of long-term datasets is essential for documenting, understanding, and forecasting such processes as the effects of climate change on ecosystems, the long-term impact of human activities (e.g., nutrient loading, stock harvesting, introduction of exotic species) on marine populations, and the formation of barriers to genetic exchange that result in speciation.

TECHNICAL FEASIBILITY AND REQUIREMENTS FOR SEAFLOOR OBSERVATORIES

Relatively capable moored-buoy and cabled observatory systems are in use today, while the more complex systems that are needed should become feasible when sufficient engineering development resources are devoted to key infrastructure elements.

CABLED SYSTEMS

Cabled seafloor observatories will use undersea telecommunications cables to supply power, communications, and command and control capabilities to scientific monitoring equipment at nodes along the cabled system. Each node can support a range of devices that might include items such as an Autonomous Underwater Vehicle (AUV) docking station. Cabled systems will be the preferred approach when power and data telemetry requirements of an observatory node are high. Early generation commercial optical undersea cable systems that are soon to be retired will have the communications capacity to satisfy most anticipated observatory research needs, but will possibly have insufficient power capability. If these cables are suitably located for seafloor observatory research studies, their use could be explored to reduce the need for expensive new cable systems.

As it is likely that cabled observatories would be installed at a site for a decade or more, substantial engineering development will be required in the design and packaging of the power conditioning, network management, and science experiment equipment. In order to meet the requirements for high system-operational time (versus downtime), low repair costs, and overall equipment lifetime, significant trade-offs will have to be considered between the use of commercially available and custom-built equipment.

MOORED BUOYS

Moored-buoy observatories consist of surface buoys acting as a central power generation and communications node, with a satellite or direct radio link to shore. This surface buoy is mechanically connected to the seafloor and communicates with instrument packages on the mooring line or on the seafloor acoustically or via an electrical or fiber optic cable. In contrast to cabled observatories, moored-buoy systems are less expensive to install, but the trade-off is a greatly diminished communications bandwidth and reduced power availability. Buoy-based observatories are well suited for long-term observations in remote areas where cabled observatories are unavailable or are prohibitively expensive to install, and for studies of episodic processes or investigations that require observations for periods of months to several years from relocatable observatories.

Ships and Remotely Operated Vehicles

Ship and Remotely Operated Vehicle (ROV) capabilities suitable for installing and maintaining seafloor observatories currently exist in industry and within the University National Oceanographic Laboratory System (UNOLS) fleet. Specialized cable-laying and support vessels will be required to install cabled and large moored-buoy systems. For operation and most maintenance purposes, the Committee believes that one to two dedicated research vessels or workboats with ROVs will be required to support an observatory system consisting of approximately two dozen nodes. Considering the current stress on ROV availability, the Committee believes that present capabilities will have to be augmented to support a major seafloor observatory program.

Scientific Instrumentation

While there are a wide variety of sensors currently available for undersea work, and there are many key instruments that are currently deployed for long time periods (such as seismometers, hydrophone arrays, and current meters), it is clear that development of new sensors will be critical for seafloor observatories to be fully effective. Sensor technology in many areas (e.g., chemistry and biology) is not sufficiently advanced to take substantial advantage of the proposed observatory infrastructure and many sensors will need considerable development before they can be expected to operate unattended for long periods of time in an observatory setting. If an ocean observatory infrastructure is to be established, a substantial parallel investment in sensor technology will be necessary. In addition, a seafloor observatory network should have the capability to incorporate visitor instruments that are dedicated to a specific experiment for a finite amount of time.

Autonomous Underwater Vehicles

AUVs provide the capability to move instruments from an observatory node to surrounding sites of interest, greatly expanding the region available for data collection. For example, AUVs can potentially undertake a variety of mapping and sampling missions while using fixed observatory nodes to recharge batteries, offload data, and receive new instructions. Furthermore, as many oceanographic processes occur episodically in relatively localized regions, it often will be necessary to search for these sporadic events over a wider area. AUVs can provide this capability. Significant engineering advances will be required for AUVs to be routinely used at seafloor observatories, including the development of a reliable docking capability at the node and the capability to operate for extended periods (a year or more) without servicing.

DATA PROCESSING, ARCHIVING, AND DISTRIBUTION

Seafloor observatories will present a great opportunity to collect long time-series datasets, but a challenge to any observatory data management structure will be the processing, distributing, and archiving of the very large datasets produced. A fully integrated plan for data handling should be developed early in the planning stages for any seafloor observatory program. This plan should include provisions for making all data publicly available as early as possible. Data management support should be provided to science investigators; in return, science proposals must anticipate and address data management issues. In addition, financial support should be made available to principal investigators for the production of data products suitable for public distribution. A distributed data management system is desirable, in order to take advantage of existing data management facilities to the extent possible.

BENEFITS AND RISKS OF ESTABLISHING A SEAFLOOR OBSERVATORY NETWORK

The establishment of a network of seafloor observatories will represent a new direction in ocean science research, and one that will require a major investment of resources over many decades. Such a major commitment carries with it both potential benefits and risks:

POTENTIAL BENEFITS

The potential benefits associated with the establishment of a seafloor observatory program include:

- establishment of a foundation for new discoveries and major advances in the ocean sciences, by providing a means to carry out fundamental research on natural and human-induced change on timescales ranging from seconds to decades;
- advances in societally relevant areas of oceanographic research, such as marine biotechnology, the ocean's role in climate change, the evalua-

tion of mineral and fishery resources, and the assessment and mitigation of natural hazards, such as earthquakes, tsunamis, and harmful algal blooms;

- improved access to oceanographic and geophysical data, enabling researchers anywhere in the world to study the oceans and earth in real-time or near real-time by providing basic observatory infrastructure with a wide variety of sensors;
- establishment of permanent observation sites over the 70 percent of Earth's surface covered by oceans, to provide truly global geophysical and oceanographic coverage not possible with observations limited to continental or island stations;
- development of new experimental approaches and observational strategies for studying the deep sea;
- enhancement of interdisciplinary research for improving the understanding of interactions between physical, biological, and chemical processes in the oceans;
- establishment of observational resources as fully funded facilities, with the use of and access to these facilities being determined by peer-reviewed proposals; and
- increased public awareness of the oceans through new educational opportunities for students at all levels, using seafloor observatories as a platform for public participation in real-time experiments.

POTENTIAL RISKS

The potential risks associated with the establishment of a seafloor observatory program include:

- installation of poorly designed and unreliable observatory systems if program and project planning and risk management are inadequate, technical expertise is lacking, and/or engineering development resources are insufficient;
- potential interference between experiments resulting from inadequate design, coordination, and/or testing of scientific instrumentation;
- inefficient use of resources if important technological questions are not adequately resolved before major investments in observatory infrastructure are made;
- possible compromise in system performance if critical technologies (e.g., satellite telemetry systems and development of some sensor types) driven by market forces outside the scientific community are not available when needed;
- the potential for a growing concentration of technical groups and expertise at a smaller number of institutions involved in supporting

the observatories, with the result that many students and scientists may become further removed from understanding how observations are made;

- unreasonable constraints on the freedom of individual investigators to choose the location and timing of their experiments;
- the potential for severe impacts on observatory science funding, and funding for other kinds of research and expeditionary science, if the cost of building, maintaining, and operating an observatory infrastructure is higher than initially estimated, and/or there is a catastrophic loss of observatory components;
- underuse of observatory infrastructure if insufficient funds are budgeted for supporting observatory-related science and the development of scientific instrumentation; and
- the potential inability of the present funding structure (based on peer-reviewed, 2- to 5-year duration, discipline-based grants) to judge the merit of projects requiring sustained time-series observations over many years or decades, and/or projects that are highly interdisciplinary.

RECOMMENDATIONS

Based on a detailed consideration of the potential benefits and risks that might be associated with a seafloor observatory program, the Committee makes the following recommendations.

1. NSF should move forward with the planning and implementation of a seafloor observatory program.

Seafloor observatories represent a promising approach for advancing basic research in the earth and ocean sciences and for addressing societally important issues. The establishment of a major seafloor observatory program will require some philosophical and intellectual reorientation within the oceanographic community, building on and complementing the more traditional focus on ship-based mapping and sampling programs. It will also require a major new commitment of resources. Based on the limited information available to the Committee, it is estimated that the initial cost of establishing a seafloor observatory infrastructure could eventually approach several hundred million dollars, and the cost of operating and maintaining this system could be several tens of millions of dollars annually. Thus, seafloor observatories may ultimately require a level of support comparable to that of the present academic research fleet. An investment of this size must be approached cautiously. In addition, mechanisms need to be put in place to deal with contingencies that arise (NRC, 1999) during the establishment of a seafloor observatory network.

2. **A comprehensive seafloor observatory program should include both cabled and moored-buoy systems. Moored-buoy systems should include both relatively high-power, high-bandwidth buoys and simpler, lower-power, limited-bandwidth buoys.**

The diverse applications for seafloor observatory science require the use of both cabled and moored-buoy observatories. Thus, the development of both systems should proceed in parallel. Because of the scientific need to study transient events, it is also important that rapidly deployable (within weeks or months) observatory systems be developed.

3. **The first step in establishing a seafloor observatory system should be the development of a detailed, comprehensive program and project implementation plan, with review by knowledgeable, independent experts. Program management should strive to incorporate the best features of previous and current large programs in the earth, ocean, and planetary sciences.**

The development of a program and project implementation plan should include a comprehensive definition of the management and science advisory structure for an observatory program, an implementation timeline and task list with specific milestones, a funding profile for the program, and a schedule for periodic review of program planning and implementation efforts by knowledgeable, independent experts. The management structure must ensure fair and equitable access to observatory infrastructure and must also provide information concerning such issues as protocols and engineering requirements for attaching instruments to a node. The requirements of a management and operational structure for a seafloor observatory program are likely to be similar to other large, coordinated programs in the earth, ocean, and planetary sciences (e.g., UNOLS, JOIDES/ODP,[1] UCAR/NCAR,[2] IRIS,[3] and NASA[4] mission structures), and the most successful features of these structures should be adopted.

[1]JOIDES - Joint Oceanographic Institutions for Deep Earth Sampling; ODP - Ocean Drilling Program

[2]UCAR - University Corporation for Atmospheric Research; NCAR - National Center for Atmospheric Research

[3]IRIS - Incorporated Research Institutions for Seismology

[4]NASA - National Aeronautic and Space Administration

4. A phased implementation strategy should be developed, with adequate prototyping and testing, before deployment of seafloor observatories on a large scale because of the cost, complexity, and technical challenges associated with the establishment of these systems.

Great care is needed in the design and implementation of seafloor observatories if they are to meet their scientific potential. Observatory networks should start with simpler nodes, adding more complex nodes and configurations over time. This growth plan will allow lessons learned from early deployments to be factored into the design of the later, more complex systems. As such, consideration should be given to the establishment of one or more pilot observatory sites to test prototype systems and sensors in areas that are readily accessible by ships and ROVs. A certified testing capability is also needed to test instrumentation and identify possible interference problems. The engineering development of some component systems that will be part of the initial pilot observatories (such as power and communications systems, AUVs, and sensor technology) could occur in parallel with the establishment of a program and project implementation plan. This will prevent delays in the establishment of initial pilot observatory needs.

5. A seafloor observatory program should include funding for three essential elements: basic observatory infrastructure, development of new sensor and AUV technology, and scientific research using seafloor observatory data.

Advances in sensor and AUV development must proceed in parallel with the development, design, manufacture, and installation of basic observatory infrastructure. The development of biological and chemical sensors and instrumentation for long-term in situ measurements is a particularly high priority. AUV development is important because these vehicles will provide the means to greatly expand the footprint of a fixed observatory node by undertaking a variety of mapping and sampling missions around the node.

There will be no benefit to the existence of seafloor observatories unless scientists are using them to advance our knowledge and understanding of the oceans. Funding for infrastructure and sensors must be balanced with funding for excellent peer-reviewed science that takes advantage of the unique capabilities of observatories. There will still be many important scientific problems that are best addressed using traditional ship-based techniques or fleets of drifters or floats. It is essential, therefore, that a seafloor observatory program be only one component of a much broader ocean research strategy.

6. **New mechanisms should be developed for the evaluation and funding of science proposals requiring sustained time-series observations over many years or decades and for proposals that are highly interdisciplinary.**

Support for observatory-related science will pose new challenges for such funding agencies as NSF. Observatory proposals will typically be more interdisciplinary and will require funding over longer time periods than is currently the norm. The NSF program and review structure is currently not well structured to handle these kinds of proposals, and NSF is taking steps to remedy this problem through increased cooperation and cooperative review of interdisciplinary proposals among program units. This deficiency was highlighted in the NRC report *Global Ocean Science: Toward an Integrated Approach*, which proposed the creation of a new unit within the Research Section of the NSF Ocean Sciences Division that would be charged with managing a broad spectrum of interdisciplinary projects (NRC, 1999).

7. **A mechanism should be developed to transition successful instrumentation developed by an individual scientist to a community asset.**

The development of mechanisms to support and encourage the transition of new instrumentation and technology from successful prototype to supported elements of the observatory infrastructure provides a major challenge both for funding agencies and the scientific community. NSF has an excellent track record of funding individual investigators to develop new instrumentation, but turning this instrumentation into a community asset has not been easy. The fundamental problem is to take an instrument that is successful in the hands of the developer and to make it successful in the hands of the broader community. Such a transition may impose significant demands beyond those an individual investigator would normally undertake, for example, re-engineering elements of the system for non-expert use, production of systems in quantity, and operational support of fielded systems. The importance of operational support can be understood by considering a straightforward Conductivity, Temperature, and Depth (CTD) capability, which requires a trained and experienced operator base despite the commercial availability of CTD hardware. Involvement of industry through existing mechanisms, such as the Small Business Innovation Research (SBIR) program, although clearly important, does not address the fundamental need of providing mechanisms for extending infrastructure support to appropriate new systems. Since the observatory can be characterized as an extension of our network and power infrastructure into the ocean, with the goal of supporting diverse oceanographic instruments, mechanisms for making such instruments widely available must be an integral part of the plan.

8. **An active public outreach and education program (including kindergarten to twelfth grade [K-12]) should be a high-priority component of a seafloor observatory program, with a specified percentage of program funding dedicated to this effort.**

Seafloor observatories with real-time communication capabilities will offer an excellent opportunity for public outreach and innovative education initiatives at all levels. With real-time data links to deep sea instruments, it will be possible to involve students and the public directly in ocean science. However, the educational and public outreach potential of seafloor observatories can only be realized by making a meaningful financial commitment to support the development of new ways to present and interpret data for the non-scientific public. To ensure that this effort is made, a percentage of program funding should be provided for outreach and every science proposal should be given a financial incentive to encourage participation in education and public outreach activities. This financial commitment should be drawn from public, private, and industry sources.

9. **A seafloor observatory program should have an open data policy, and resources should be committed to support information centers for archiving observatory data, generating useable data products, and disseminating information to the general public.**

Routine (facility-generated) information products and data should be archived in a central facility and be made available in as near real-time as possible to all investigators and the general public, ideally through the Internet. Data from experimental sensors or individual investigator-initiated experiments should be made publicly available after quality control procedures have been applied, but within 1 to 2 years of retrieval. A distributed data management system is desirable and, to the extent possible, existing data management facilities should be used (NRC, 1999).

10. **Seafloor observatory programs in the United States should be coordinated with similar international efforts to the extent that progress in the U.S. program is not inhibited. In addition, the Committee recommends that the potential of an integrated, international observatory program be explored.**

The goals of a seafloor observatory program in the United States are closely linked to a number of ongoing national and international initiatives, such as GEOSCOPE and the International Ocean Network. Where practical, coordination of these efforts at the international level will be beneficial and in some cases essential. Some major scientific objectives (e.g., global seismic coverage) will be achievable only through this kind of global cooperation.

The vision of establishing a global network of seafloor observatories holds tremendous promise for advancing our understanding of Earth and its oceans. The Committee recognizes that realizing this vision will be difficult and expensive, but based on examination of the important scientific questions that remain to be answered and the current state of technology, the Committee believes the time has come to take the first concrete steps.

Introduction

CHALLENGES AND OPPORTUNITIES

The ocean exerts a pervasive influence on Earth's environment; thus, it is important that we learn the intricacies of how this system operates (NRC, 1992; 1998b; 1999). For example, the ocean is an important regulator of climate change through its surface temperature and its control on atmospheric composition (IPCC, 1995). Understanding the link between natural and anthropogenic climate change and ocean circulation is essential if we are to predict the magnitude and impact of future changes in Earth's climate. Understanding our ocean and the complex physical, biological, chemical, and geological systems operating within it should be an important goal for the coming decades of the 21st century. This increased understanding will capture the public's imagination, and it will also, by necessity, bring about advancements in fields as diverse as engineering and biomedical technology. Another important motivating force to increase our understanding of ocean systems is that the global economy is highly dependent on the ocean for tourism, transportation, fisheries, hydrocarbons, and mineral resources (Summerhayes, 1996, and references therein). Thus, with continued population growth, it is inevitable that ocean resources will be increasingly relied on and will need to be used more effectively.

Recent oceanographic research has demonstrated that to understand the complex interaction of the various ocean systems, long-term time-series measurements of critical oceanographic parameters are needed to supplement traditional seagoing investigations (NRC, 1999). The more traditional method of investigating components of the ocean environment in isolation must be incorporated into a more holistic approach. The establishment of a global network of seafloor observatories will help provide the means to accomplish this goal.

In this report, seafloor observatories are defined as unmanned, fixed systems of instruments, sensors, and command modules connected either acoustically or via a seafloor junction box to a surface buoy or a fiber optic cable to land. These observatories will have power and communication capabilities and will provide support for spatially distributed sensing systems and mobile platforms. Sensors and instruments that are used at seafloor observatories will potentially collect data from above the air-sea interface to below the seafloor and will provide support for in situ manipulative experiments. Seafloor observatories will also be a powerful complement to satellite measurement systems by providing the ability to collect vertical measurements within the water column for use with the spatial measurements acquired by satellites while also providing the capability to calibrate remotely sensed satellite measurements.

Ocean observatory science has already had major successes. For example, the Tropical Atmosphere-Ocean (TAO) array has enabled improved detection, understanding, and prediction of El Niño events and is an example of the achievements that can be accomplished with simple systems. TAO consists of approximately 70 moored ocean buoys in the tropical Pacific Ocean that telemeter oceanographic and meteorological data to shore in real-time via the ARGOS satellite system. TAO Autonomous Temperature Line Acquisition System (ATLAS) buoys (Figures 1-1 and 1-2) measure surface winds, air temperature and relative humidity, and ocean temperatures in the upper 500 m of the ocean, whereas TAO Equatorial Current Meter buoys include additional instruments to measure ocean currents and variables, such as shortwave radiation and rainfall.

Another success is the SOund SUrveillance System (SOSUS), which is a fixed component of the U.S. Navy's Integrated Undersea Surveillance Systems network used for deep-ocean surveillance during the Cold War. SOSUS consists of bottom-mounted hydrophone arrays connected by undersea communication cables to facilities on shore. The combination of location within the oceanic sound channel and the sensitivity of large-aperture hydrophone arrays allows the system to detect radiated acoustic power of less than a watt at ranges of several hundred kilometers. SOSUS is an important tool for both continuous monitoring of low-level seismicity around the northeast Pacific Ocean and real-time detection of volcanic activity along the northeast Pacific spreading centers, and has provided a useful means to track whale migrations (Plate I).

Although traditional seagoing investigations will continue to be prominent in oceanographic research, the question posed to this Committee was whether there is scientific justification for the establishment of a major coordinated seafloor observatory effort and whether such an effort is technically

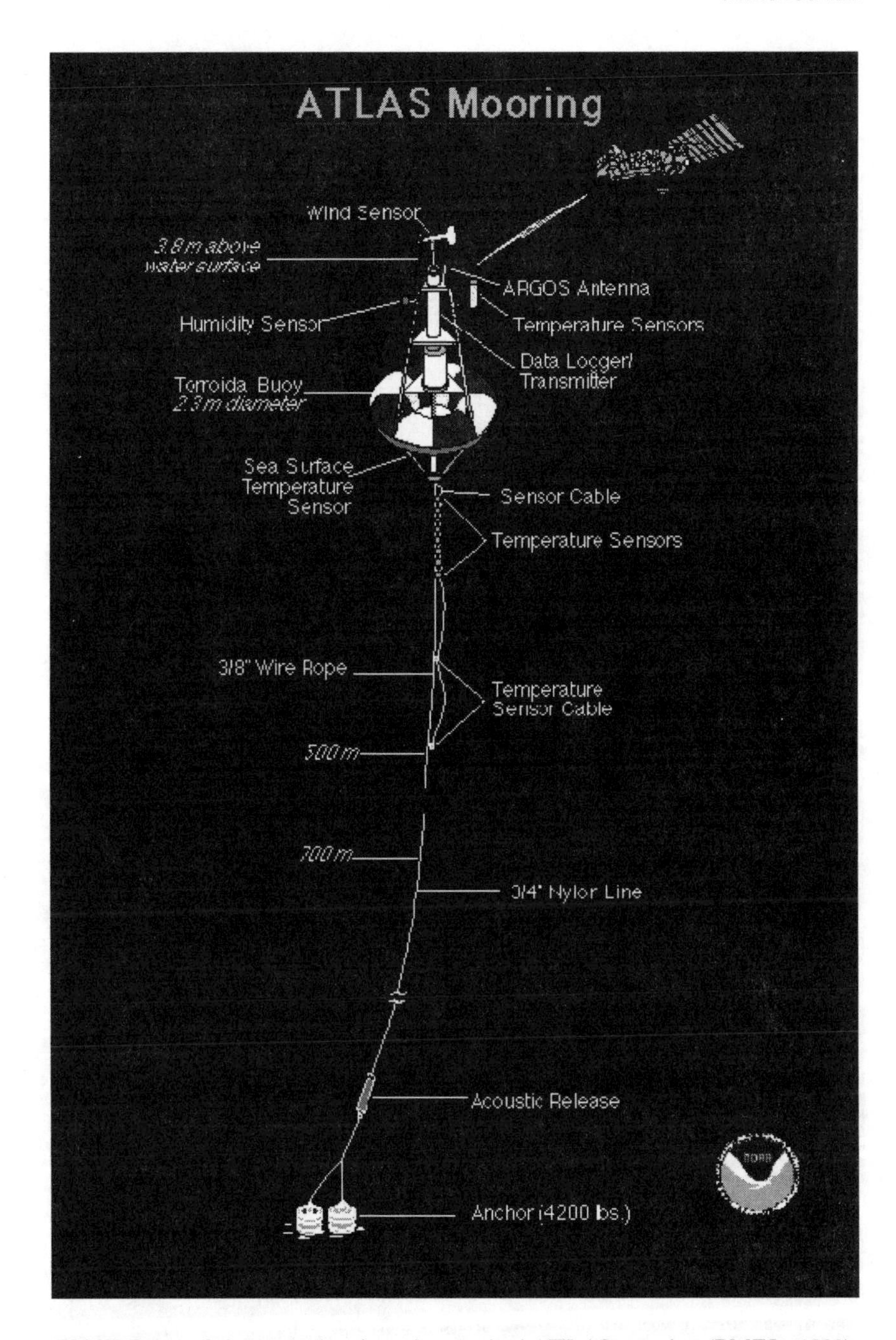

FIGURE 1-1 Schematic drawing of a standard ATLAS mooring (PMEL, 1999).

FIGURE 1-2 Recovering an ATLAS mooring. The ATLAS buoy consists of a toroidally shaped fiberglass casing over foam with an aluminum tower and a stainless steel bridle. The toroid is 2.3 m in diameter. When completely rigged, the system has an air weight of approximately 225 kg, a net buoyancy of nearly 1800 kg, and an overall height of 4.9 m. The electronics tube is approximately 1.5 m long, 0.18 m in diameter, and weighs 27 kg. The buoy can be seen on radar for 4 to 8 miles depending on sea conditions (PMEL, 1999).

Box 1-1
STATEMENT OF TASK

This study will examine the scientific merit of, technical requirements for, and overall feasibility of establishing the infrastructure needed to implement a system of seafloor observatories. Recently, many seafloor observatory programs have been discussed or proposed. This study will assess the extent to which seafloor observatories will address future requirements for conducting multi-disciplinary ocean research, and attempt to gauge the level of support for such programs within ocean science and the broader scientific community.

feasible in the near future. To this end, the National Science Foundation (NSF) requested a study from the National Research Council's (NRC) Ocean Studies Board (OSB) to investigate the scientific merit and technical feasibility of establishing a series of seafloor observatories. A steering committee of eight members representing major areas of oceanographic science and ocean engineering was appointed to addresses the Statement of Task (Box 1-1) and write this externally reviewed consensus report. Although the statement of task for this study states that the committee will "gauge the level of support for such programs within ocean science and the broader scientific community," no statistical measures were used. Instead, the very positive response to the symposium over a broad range of scientific disciplines and the enthusiastic nature of the discussions that occurred were used as evidence for broad community support for this initiative.

WHY ESTABLISH SEAFLOOR OBSERVATORIES?

In recent decades ocean, earth, and planetary sciences have been shifting from an intermittent, expeditionary mode of exploration and problem definition toward a mode of sustained in situ observation and experimentation. The motivation for this change is intellectually grounded and stimulated by new scientific discoveries and the unanticipated effects of earth and ocean processes on mankind. Examples of these include:

- the detection of formerly unknown, chemosynthetically based ecosystems hosted on and beneath the seafloor;

- the widespread destruction wrought by the Northridge, California, and Kobe, Japan, earthquakes;
- the intense influence on global weather of the El Niño events of 1982-1983 and 1997-1998; and
- the clear realization that humans have affected marine ecosystems worldwide, both directly and indirectly through release of organic and inorganic pollutants, harvesting of fish and shellfish, and introduction of exotic species, and indirectly through climate alteration.

This change in the mode of investigation stems from the realization that Earth and its oceans are not static, but are dynamic on many time and space scales, not just the relatively short timescales involved in the catastrophic examples above. Thus, understanding Earth and its oceans requires investigating processes as they occur, which cannot be satisfactorily accomplished with occasional mapping and sampling. A scientifically powerful component of the observatory concept is the collection of multiple oceanographic variables at a single location. These multidisciplinary datasets will enable the enhancement of more traditional oceanographic methods and allow for the development of new and creative ways of doing ocean science. While spatial mapping and exploration remains essential in setting the stage for process-oriented studies, a sustained time-series approach will be required to truly comprehend earth and ocean processes, and to develop predictive capabilities. Such an approach is implicit in recommendations of the recent NSF long-range planning "Futures" reports (Baker and McNutt, 1996; Jumars and Hay, 1999; Mayer and Druffel, 1999; Royer and Young, 1999). For example, the APROPOS[1] Report states that "long time-series observations . . . provide essential data on oceanographic processes, particularly those related to climate change" and recommends "a national effort to support sustained high-quality global observations over decades" (Royer and Young, 1999).

In essence, this approach requires a sustained observational presence on the seafloor. In some subdisciplines, the seafloor observatory initiative is taking the form of adaptive "observatory" science based on long-term deployment of a wide variety of seafloor instrumentation, occasionally with real-time data transmission and instrument control. A strong case has been made that such capabilities will offer earth and ocean scientists unique potential to (a) study multiple interrelated processes over a range of timescales, in some cases conducting in situ perturbation experiments involving artificial modification of the natural environment to observe the effects of this modification; (b) conduct comparative studies of regional processes and spatial variability;

[1]APROPOS - Advances and Primary Research Opportunities in Physical Oceanography Studies

and (c) map whole-earth and global ocean structures tomographically[2] using artificial and natural source signals.

Fulfilling the scientific and societal potential of the seafloor observatory approach will entail significant philosophical and intellectual re-orientation within the oceanographic community, building on and complementing the more traditional focus in some disciplines on short-term spatial mapping and sampling. The reorientation for other disciplines with longer histories of collecting sustained time-series datasets will be more technological, particularly in terms of automating recording and also using two-way real-time communication.

TYPES OF SEAFLOOR OBSERVATORIES DISCUSSED IN THIS REPORT

In this report a "seafloor observatory" is defined as an unmanned system at a fixed site in the ocean providing power, command and control, and communications to sensors located on or below the seafloor, in the overlying water column, or at the air-sea interface. Sensors may be acoustically, electrically, or fiber-optically linked to the observatory node. The node may also support long-endurance mobile vehicles (e.g., Autonomous Underwater Vehicles [AUVs] and Remotely Operated Vehicles [ROVs]) that are capable of repeat surveys of a broader area around each node. We do not include various Lagrangian[3] drifters or floats in this definition, but recognize that these systems would complement an array of fixed observatory sites as part of an integrated ocean observing system. In this report, we consider two classes of seafloor observatories: cable-based observatory networks and moored-buoy observatories.

Cabled-based observatories will use undersea telecommunications cables to supply power, communications, and command and control capabilities to scientific monitoring equipment at nodes along the cabled system. Each node can support a range of devices that may include equipment such as an AUV docking station. Cabled systems will be the preferred approach when power and data telemetry requirements of an observatory node are high. The high cost of fiber optic cable and the need for a given location to support a nearby cable landing will limit the spatial coverage of these observatories.

Moored-buoy observatories consist of a surface buoy acting as a central instrument node with a satellite or direct radio link to shore. The surface buoy is connected to the seafloor node acoustically or via an electrical or fiber optic

---

[2]Tomographically - the use of changes in sound velocity to map variations in ocean structure or water mass circulation.

[3]Lagrangian - following the path taken by a parcel of water as it moves relative to the earth.

riser, or both. Instruments and observatory devices are either directly connected to the seafloor node or can communicate via an acoustic communication link. Power generation for moored-buoy observatories can be achieved by a variety of means depending on the required wattage. These include solar and wind for low-power requirements and diesel generators for high-power requirements. Mooring-based observatories would be well suited to meet the needs of several major areas of science activities, including (1) studies of episodic events, (2) process studies for periods of months to several years, and (3) long-term observation in remote areas where cabled observatories are unavailable or prohibitively expensive to install. The spatial coverage of moored buoys will be dependent on their complexity and thus their unit cost. In general, moored buoys are lower in cost than for the equivalent capability in a cabled observatory. As such, buoys can be deployed in greater number and act as a network of nodes that have the potential to provide significant areal coverage.

MAJOR PROPOSED AND ACTIVE OBSERVATORY PROGRAMS

The scientific benefits of establishing a seafloor observatory network for oceanographic, climatic, and meteorological investigations have been recognized for many years. As such, numerous independent national and international observatory efforts have been proposed or are underway. Although a description of all of these observatory efforts is beyond the scope of this report, some are summarized here and in the chapters that follow.

During the 1990s several workshops were held outlining long-term scientific strategies for the use of observatories in geophysics research. In 1996, a subset of the convenors for these workshops met to discuss the establishment of a national seafloor observatory initiative, concentrating on geophysics, that would combine but not subsume existing individual observatory efforts. In 1997, the initiative was named Deep Earth Observatories on the Seafloor (DEOS) and its membership and charter were formalized. Initially, the focus of DEOS was on deep-water geo-observatories, but this subsequently was expanded to include nearshore observatories and water-column studies. To reflect this effort to engage the wider oceanographic community, the acronym definition was changed to Dynamics of Earth and Ocean Systems in 1999. DEOS steering committee members represented the major national geoscience observatory programs (BOREHOLE,[4] CABLE, RIDGE,[5] MARGINS, and OSN[6]) with additional membership from the microbiological community (University of Miami, 1999).

[4]BOREHOLE - BOREHole Observatories, Laboratories, and Experiments.

[5]RIDGE - Ridge InterDisciplinary Global Experiments.

[6]OSN - Ocean Seismic Network.

One major component of the DEOS planning effort is North East Pacific Time-series Undersea Networked Experiments (NEPTUNE), a proposal to establish a plate-scale observatory network on the Juan de Fuca Plate off the coasts of Oregon and Washington linked to land-based research laboratories and classrooms using high-speed, fiber-optic submarine telecommunication cables (NEPTUNE, 2000). This system has been designed to provide real-time data transmission to shore, interactive control over robotic vehicles on site, and power to instruments and vehicles. As many globally significant earth processes operate at or below the scale of tectonic plates, the proposed rationale behind NEPTUNE is that the seafloor observatory network should be constructed at the scale of a lithospheric plate. The site selected is the Juan de Fuca Plate located within a few hundred kilometers of the U.S.-Canadian west coast. It is proposed that NEPTUNE will allow scientists and educators to analyze and use data bearing on the linkages between key oceanographic and plate tectonic processes (NEPTUNE, 2000).

Another major component of DEOS is the establishment of a permanent OSN consisting of ~20 sites throughout the world's oceans for improved geophysical imaging of the internal structure of Earth. These planetary-scale, fixed ocean observatories may also serve as long-term measurement sites for other types of oceanographic and climate studies, such as those envisioned by the Global Eulerian Observations (GEO) system of moorings (Plate II). The goal of GEO is to establish oceanographic observatories at select sites around the world's oceans for the collection of time-series measurements of surface meteorology; air-sea exchanges of heat, freshwater, and momentum; and full-depth profiles of water properties, including temperature and salinity, and ocean velocity. It is proposed that recent and future advances in mooring and instrumentation technology will enable the maintenance of the GEO observatories at a fraction of the cost of previous ocean weather stations. If successful, the sites will be an important element in an integrated observational system by providing the data necessary to develop a description of the ocean's role in climate. These observatories will also provide key observations of water mass formation and transformation. In addition, the data collected could be used to quantify the transports of the major ocean current systems, to assess vertical variability in ocean structure, and to document the role of eddy processes in the transport of heat and other properties. Time-series measurements from the GEO observatories will be an essential element of the strategy developed to construct accurate fields of air-sea fluxes. These observation stations have been proposed as an important component of the Global Ocean Observing System (GOOS).

GOOS is an observatory framework formally initiated by the Intergovernmental Oceanographic Commission Executive Council in 1992 in cooperation with the World Meteorological Organization, United Nations Environment Programme, and the International Council of Scientific Unions (GOOS Project

Office, 1999; NRC, 1997). GOOS was conceived as an international system for gathering, coordinating, quality controlling, and distributing oceanographic data and derived data products as defined by the requirements of user groups. Support for planning and implementation is apportioned among GOOS sponsoring organizations and is supplemented through these organizations by financial and in-kind contributions from participating nations. The implementation is largely dependent on the commitment of supporting nations to their national observing systems. This commitment not only includes the infrastructure of the observatories themselves, but also the scientific and technical research that is needed to support data centers.

The purpose of GOOS is to provide a framework to ensure long-term, systematic observations of the global ocean and to provide the mechanisms and infrastructure to make these data available to various nations for the solution of problems related to environmental change. GOOS is organized to resemble the global meteorological observation and prediction network presently supported by individual nations and implemented through the contributions of national agencies, organizations, and industries. Thus, observatories established under the auspices of GOOS will be primarily operational in nature. GOOS will include existing observing systems in addition to new systems, such as the proposed GEO observatories discussed above, that may become part of the GOOS network. Furthermore, it is likely that much of the data collected from a network of research-driven observatories, such as that proposed here, would be merged into the datasets that will be collected as part of GOOS. Much of the impetus for the GOOS plan has come from the need for operational oceanographic data to improve nowcasts and forecasts of ocean conditions and weather and climate.

With the numerous multipurpose ocean observatory efforts in place or proposed, some of which are discussed in boxes later in this report, a great opportunity exists for synergy among these groups. This interaction was an important point of discussion at the symposium, and numerous common areas of interest were noted. For example, there is significant overlap in the locations needed to complete the OSN and those sites necessary for the GEO observatories (Plate II). If a major network of seafloor observatories were planned in the future, to conserve resources and share common technology it will be important to develop channels for interaction between established and proposed observatory efforts.

It was not part of the Committee's charge to assess the strengths and weaknesses of existing observatory programs. A thorough review of these programs will be an important task during the drafting of an implementation plan for establishing a seafloor observatory program. When writing this implementation plan, the design strengths and weaknesses of existing programs should be considered in detail.

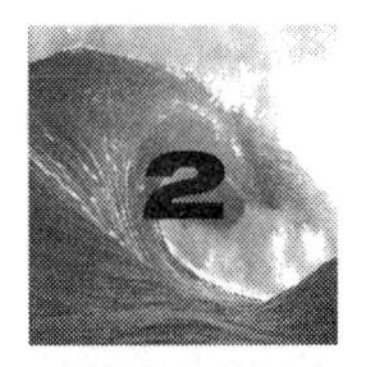

The Scientific Potential of Seafloor Observatories

The varied scientific presentations at the workshop demonstrated the significant potential of observatory science. Areas where time-series data collected from seafloor observatories would advance research include:

- *Studies of episodic processes*—Episodic processes include eruptions at mid-ocean ridges, deep-ocean convection at high latitudes, earthquake swarms at subduction megathrusts, and biological, chemical, and physical impacts of episodic storm events. These processes can be anticipated only in a statistical sense; accurate prediction of the timing for an individual event is not possible. Lower-complexity mooring-based observatories are beneficial for the study of episodic processes, as they provide a lower-cost system for deployment in anticipation of or in the aftermath of an episodic event.
- *Process studies for periods of months to several years*—The ability to deploy an observatory in remote locations to investigate oceanographic processes occurring on timescales of months to years is a critical need of the oceanographic community. These processes include hydrothermal activity and biomass variability in vent communities along portions of the mid-ocean ridge system, air-sea interactions in the Southern Ocean, and biological and chemical variability of the water column at both coastal and oceanic sites.
- *Observations of global and long-term processes*—Moored observatories will be essential for the establishment of an observatory network to investigate global processes, such as the dynamics of oceanic lithosphere and thermohaline circulation. These moored systems would also

provide long-term capabilities in remote regions where cabled observatories are unavailable or would be prohibitively expensive to install.

The sections in this chapter summarize the discussions and scientific outcomes of breakout groups organized according to the six National Science Foundation (NSF) decadal report themes based on the "Futures" workshops (Baker and McNutt, 1996; Jumars and Hay, 1999; Mayer and Druffel, 1999; Royer and Young, 1999). Each section addresses future directions and major scientific problems, the role of sustained time-series observations, and technical requirements for seafloor observatories. Boxes describing currently active observatory or time-series experiments have been included where they are appropriate to the science being discussed. In addition, tables (developed based on symposium discussions) have been added to each section, highlighting areas where observatories are "very useful" in investigating a scientific problem, and where they are "useful." The term "very useful" is used to categorize scientific problems for which long-term, time-series datasets collected at seafloor observatories will result in substantial scientific progress that will either not be possible using a traditional expeditionary approach or can only be accomplished with limited success using more traditional means. "Useful" is used to categorize those scientific problems for which an observatory approach will provide a valuable complement to other more traditional research strategies.

ROLE OF THE OCEAN IN CLIMATE

Climate variations have widespread societal, economic, and environmental impacts. As a result, vigorous research efforts are currently aimed at improving understanding of the spectrum of climate system variations, discovering potentially predictable elements, and exploiting that potential. For example, El Niño/Southern Oscillation (ENSO) forecasts have demonstrated both the potential and the value of climate prediction.

The ocean is an intrinsic component of Earth's climate system, playing an increasingly important role in determining the nature of climate variability as timescales increase. Researchers seek to improve our understanding of the role of ocean heat storage, transport, and release in the coupled ocean-atmosphere-land climate system, and of the interactions of oceanic biogeochemistry with the climate system (Boxes 2-1 and 2-2).

Future Directions and Major Scientific Problems

Ultimately, we seek to predict climate variability and change. This requires accurately predicting the evolution of the ocean (especially near-surface temperatures) when changes in atmospheric forcing occur. Despite impressive progress with forecasting seasonal-to-interannual evolution associated with ENSO events, there is considerable room in these and longer timescales for improvement in the observation, analysis, and assessment of predictability of the oceanic component of the coupled climate system. For example, it is now clear that Pacific decadal climate variability influences ENSO, and it is possibly strongly coupled to ENSO. Research has conclusively shown that the ocean exerts an important influence on decadal climate variability, but details of its role are unclear. Several hypotheses have been developed to explain the sparse oceanographic observations and climate model behavior. Another example involves the North Atlantic Oscillation influencing ocean physics on decadal time scales, such as Labrador Sea wintertime deep convection and the path of the Gulf Stream extension. Limited observations and models suggest that there is a feedback of the resultant ocean variability onto the atmosphere.

An important problematic challenge before us is separating natural interannual-to-centennial climate variations from anthropogenically induced climate change. This understanding is critical for predicting future variations and magnitudes of climactic change.

For both hypothesis testing and prediction purposes, we rely increasingly on models of the climate system. Even though the present generation of ocean general circulation models are much improved, representation of important ocean physics is still crude. A substantially improved observational basis for determining the needed model enhancements is required. Specific ocean science challenges include quantifying and understanding turbulent mixing; convection; water-mass formation and destruction; thermohaline circulation and its coupling to the wind-driven circulation; the generation, maintenance, and destruction of climatic anomalies; climatic oscillations and the extratropical coupling of the ocean and atmosphere on seasonal, decadal, and interdecadal timescales; and the physics of exchange processes between the ocean and the atmosphere (Royer and Young, 1999).

The U.S. scientific community has played an important role in shaping and is actively participating in the World Climate Research Programme's Climate Variability and Predictability (CLIVAR) program in order to address the challenges outlined above (NRC, 1994a; NRC, 1996a; NRC, 1998b; NRC, 1998c). CLIVAR scientists have worked with other ocean programs to develop plans for a global ocean-observing system that will meet the research requirements for major advances in understanding the role of ocean processes in the climate system and will provide a basis for predictability research (NRC, 1994b; NRC, 1997).

THE ROLE OF SUSTAINED TIME-SERIES OBSERVATIONS

Oceanographic variability has a significant influence on climate. Because of nonlinear scale interactions, it is essential for the study of climate to fully

BOX 2-1
THE HAWAII OCEAN TIME-SERIES

Objectives: The Hawaii Ocean Time-series (HOT) objectives are to document seasonal and interannual variability of water masses in the North Pacific Ocean subtropical gyre; to relate water mass variations to gyre fluctuations; to develop a climatology of short-term variability; to document and understand seasonal and interannual variability in the rates of primary production, new production, and particle export from the surface ocean; to determine the mechanisms and rates of nutrient input and recycling, especially for nitrogen and phosphorus in the upper 200 m of the water column; to measure the time-varying concentrations of dissolved inorganic carbon in the upper water column; and to estimate the annual air-to-sea carbon dioxide flux.

To achieve these objectives, biogeochemical and physical observations are collected monthly or near-monthly during three-day shipboard occupations of Station ALOHA. A subsurface sediment trap mooring is maintained at ALOHA, along with a surface mooring with meteorological, bio-optical, and physical sensors approximately 40 km away. A coastal station (Kahe Point) is also sampled during most cruises. Information is provided in Karl and Lukas (1996). Additional information on the HOT time-series can be found at http://hahana.soest.hawaii.edu/hot/hot.html (HOT, 2000).

Location: Station ALOHA is located 100 km north of Oahu in 4,740 m of water. The Kahe Point station is about 10 km offshore in water depth of about 1,500 m (Figure 2-1).

Established: HOT was established in late 1988 with National Science Foundation funding under the auspices of the Joint Global Ocean Flux Study (JGOFS) and World Ocean Circulation Experiment (WOCE) programs. The surface mooring was first deployed in early 1997.

resolve many scales of variability, and the consensus is that this will require nested, complementary observing systems (Table 2-1). In particular, the vision that is developing for an ocean-observing system for climate consists of a number of highly instrumented fixed sites around the world, a larger number of fixed sites of lesser capability, and a global array of Lagrangian samplers, all working in concert with global satellite remote sensing and global data assimi-

BOX 2-1 CONTINUED

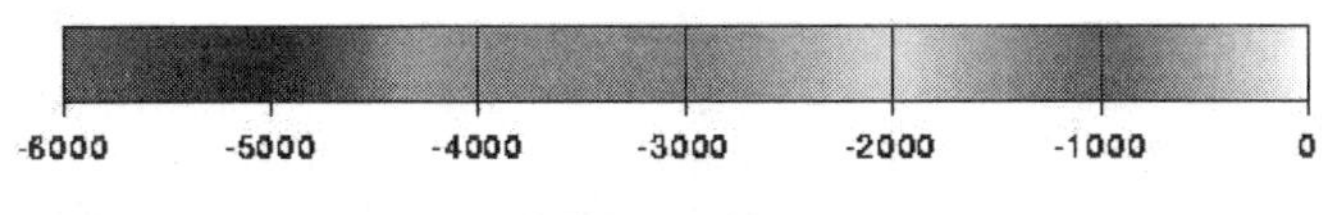

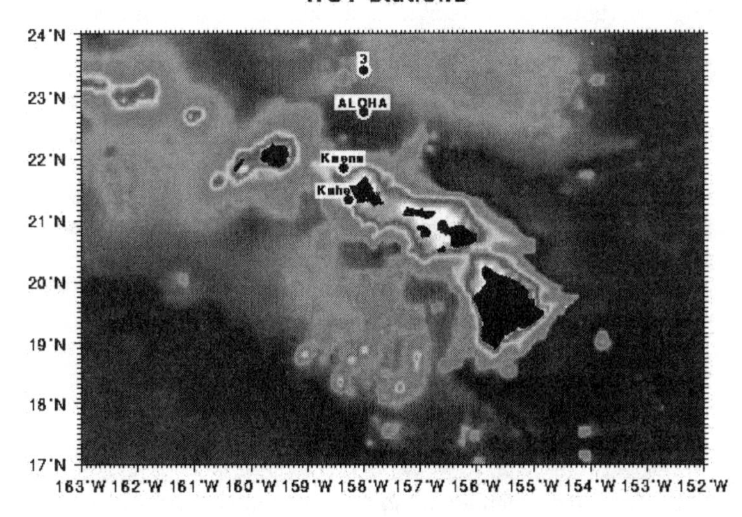

FIGURE 2-1 Locations of HOT sampling stations with water depth indicated by the grayscale defined above (Lukas and Karl, 1998).

lative modeling. Fixed site observatories would primarily consist of moorings, which are best suited for vertical and temporal sampling. Moorings also provide the means to sample scientifically critical regions, such as the upper few tens of meters of the ocean under ice cover in confined current systems and in abyssal layers (including bottom boundary layers; see "Turbulent Mixing and Biophysical Interaction" section in this chapter).

BOX 2-2
BERMUDA ATLANTIC TIME-SERIES STUDY

Objective: The Bermuda Atlantic Time-series Study (BATS) commenced monthly sampling in the western North Atlantic subtropical gyre as part of the U.S. Joint Global Ocean Flux Study (JGOFS) program in 1988. The goals of U.S. JGOFS time-series research are to better understand basic processes controlling ocean biogeochemistry on seasonal to decadal timescales, to determine the role of the oceans in the global carbon budget, and ultimately to improve our ability to predict the effects of climate change on ecosystems.

The BATS uses a monthly shipboard sampling scheme to resolve seasonal patterns and interannual variability. Core cruises last four to five days during which hydrography, nutrients, particle flux, pigments and primary production, bacterioplankton abundance and production, and complementary ancillary measurements are made. This study also incorporates data from nearby Hydrostation S, the Ocean Flux Project, and the Bermuda Testbed Mooring (BTM). Hydrostation S, established in 1954, is one of the longest-running oceanic and atmospheric time series. All of the data from the BATS program and many of the data from the Hydrostation S and BTM are available publicly at the Bermuda Biological Station for Research home page at http://www.bbsr.edu (and then following the links to the BATS data).

Location: The BATS station lies 82 km southeast of the island of Bermuda (31° 40' N, 64° 10' W) in the Sargasso Sea, approximately 1,200 km from the east coast of the United States (Figure 2-2). Bottom depth at the BATS deployment area is ~4,680 m.

Established: Monthly sampling commenced in October 1988.

In addition, observatories, together with other elements of the nested observing system (such as Lagrangian samplers and satellites), provide the physical oceanographic context for interpreting biological and chemical distributions.

Considerable planning for observatory-based science necessary for climate research has been conducted by the oceanographic community. The Ocean Observing System Development Panel (OOSDP) developed the scientific basis for global ocean observations for climate and prioritized the observations that would be made (Nowlin, 1999). The Ocean Observing Panel for Climate

BOX 2-2 CONTINUED

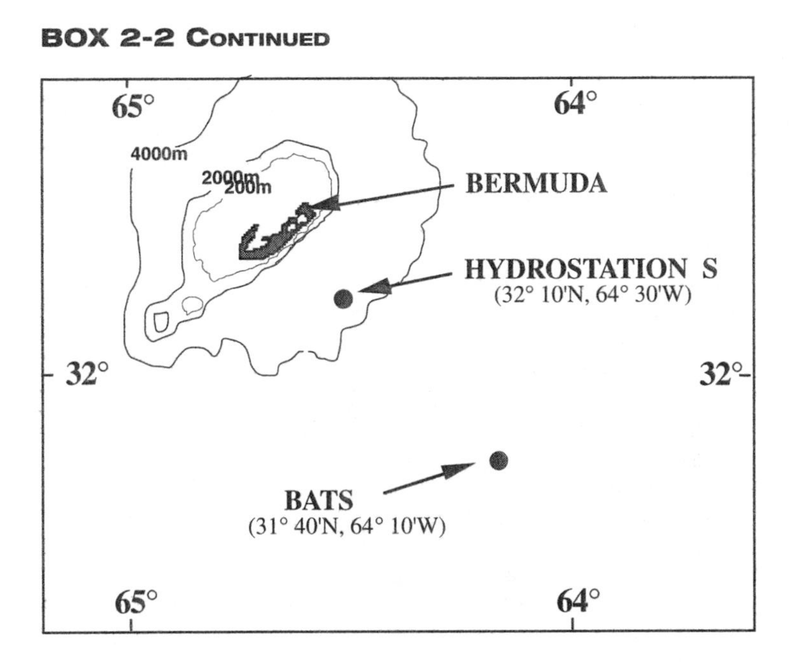

FIGURE 2-2 Location of the BATS sampling station. SOURCE: Deborah K. Steinberg, Bermuda Biological Station for Research.

(OOPC) is moving toward implementation of the OOSDP recommendations. OOPC and other groups representing climate research interests hosted a major international conference, OceanObs99, in St. Raphael, France, in October 1999 (OCEANOBS99, 1999); the conference gathered broad input from the ocean science and climate communities and helped further the scientific rationale for sustained observations.

At present, we have very few sustained observing sites supporting research on the role of the ocean in climate. To further our understanding of the role of

TABLE 2-1 The Role of Ocean in Climate: Areas Where Observatories Are Very Useful to Investigate a Particular Scientific Problem and Where They Are Useful

Observatory science is VERY USEFUL in accomplishing the following:

- Test and improve ocean circulation models;
- Observe and understand extratropical coupling of ocean and atmosphere on seasonal to interdecadal timescales;
- Understand the physics of the exchange processes between the ocean and atmosphere;
- Observe the generation, maintenance, and destruction of ocean climate anomalies;
- Predict climate variability and change;
- Monitor, understand, and predict
  - the sequestration of carbon dioxide in the ocean;
  - productivity and biomass variability, including identification of the factors that control them;
  - the full temporal and vertical evolution of thermohaline structure;
  - rapid episodic changes of the ocean (e.g., mixed-layer response to hurricanes, deep convection, meridional overturning circulation);
  - changes in water mass transformation processes;
  - air-sea exchanges of heat, moisture, momentum, and gases;
  - thermohaline variability in the Arctic and Antarctic;
  - vertical exchanges of heat, salt, nutrients, and carbon;
  - the pathways of ocean transports, such as deep western boundary currents; and
  - the role of eddies in transport and mixing.
- Provide reference sites for calibration or verification of
  - air-sea fluxes from numerical weather prediction models, satellites, and other methods;
  - absolute interior and Ekman layer velocities;
  - remotely sensed variables (sea surface temperature, sea level, wind, color); and
  - model statistics, physics, and parameterizations and how they change in evolving climate systems.

Observatory science is USEFUL in investigating the following:

- Water mass formation and destruction; and
- The relationship of heat and freshwater fluxes to wind and buoyancy forcing.

the ocean in climate, seafloor observatories should be long-term (outliving individual investigator proposals), shared-use facilities that sustain climate community measurement goals while allowing technology to evolve. A fundamental change achieved by pursuing the observatory concept would be maintenance of existing sites and establishment of new sites. This may be the key to moving from our current focus on long-term science projects, such as CLIVAR (which may result in 5- to 10-year time series), to the implementation of a sustained global ocean-observing system.

TECHNICAL REQUIREMENTS

The technical requirements of the climate and turbulent mixing communities have an important overlap, and there are compatible needs for platforms. The strategy of nested, complementary elements of an observing system is shared, and it is agreed that fixed observatories are uniquely able to support the required observations.

For the climate-science issues discussed above, the objectives do not, in general, require high power (present systems work at 5 W and less) or high data rates (tens to hundreds of numbers are now sent back per hour or per day). Long-range acoustic tomography requires more power (approximately 200 W) for driving acoustic sources. Acoustic tomography also produces large volumes of data that are not presently relayed using real-time telemetry.

The availability of ocean observatory sites around the globe that could provide power and access to data telemetry and two-way communication would enable the climate and mixing communities to make greater progress toward their science goals. Furthermore, at the symposium, it was noted that the map of Global Eulerian Observatory (GEO) sites produced by the OceanObs99 climate conference had a number of sites close to sites proposed for the Ocean Seismic Network (OSN) (Plate II). This synergy emphasizes the need for a dialog between proponents of observatories for different science goals to maximize mutual benefits and to explore how observatory sites occupied for other reasons could enable further progress in climate and mixing research.

Sites shown on Plate II fall into the category of global observatories, and would be key elements of the nested, complementary network required for global ocean observations for climate studies. These are viewed as relatively high-power, high-capability sites that would provide opportunities for instrument integration, data access, and a surface expression. Their design should minimize radio frequency contamination of sensors, shadowing of sunlight, disturbance of flow in the atmosphere and ocean, and biological and chemical contamination of the ocean and atmosphere. The mooring design should also permit the storage and release on demand of garage floats, autonomous underwater vehicles (AUVs), weather balloons, and other autonomous instrument packages. These high-capacity sites would become focal points for multidisciplinary oceanographic science and should contain a standard suite of instrumentation including bottom-pressure sensors; inverted echo sounders; electric-field sensors; temperature, salinity, velocity, surface meteorological, optical, and aerosol systems; and sensors to measure key chemical, optical, and biological parameters.

Relocatable observatories consisting of one or more moored platforms would be highly valuable for climate-related process studies involving turbulent mixing, air-sea exchange, biochemistry, coupled upper-ocean optics, etc. Different types of relocatable observatories should be considered for investigating specific science questions. These could be repositioned for one or more years at different locations around the globe as needed, such as proposed for surveying turbulence parameter space in the "Technical Requirements" of the "Turbulent Mixing and Biophysical Interaction" section. Relocatable observatories should have power available along with data transmission capabilities. There was interest expressed at the symposium in taking relocatable observatories to extreme environments (for example, areas with high wind speeds) and data-sparse regions where significant impact on climate science would be anticipated. Further development of moored turbulence sensors is needed for use on relocatable observatories.

Permanent cabled observatories are of interest as sites where multidisciplinary "laboratories" could be developed. Of particular interest would be efforts to fully image a three-dimensional (3-D) volume in the ocean, resolving the time and space variability of the physics, biology, chemistry, and geology on scales from centimeters to kilometers (the 3-D aspect is discussed in more detail in the "Turbulent Mixing and Biophysical Interaction" section). Such observatories should have the physical infrastructure to permit growth and easy addition of new users and instruments.

FLUIDS AND LIFE IN THE OCEANIC CRUST

The chemistry and biology of fluids within the oceanic crust is a cutting-edge research field for which seafloor observatories are thought to be a needed investigative approach (Table 2-2). One of the most exciting scientific problems that can be addressed using observatory science concerns the nature of the subsurface biosphere, thought to contain a population of dormant microbes that are periodically driven into a population explosion by input of heat and volatiles into the crust during magma emplacement events (Boxes 2-3 and 2-4; Delaney et al., 1998; Summit and Baross, 1998). In addition to the ridge crest and flanks, a population of microbes is also thought to exist in the extreme environment of subduction zones sustained by either continuous or episodic input of energy and nutrients (Cragg et al., 1995). Another exciting and closely related scientific problem concerns the general response of the hydrothermal system and associated biota to seafloor spreading events in which magma is

TABLE 2-2 Fluids and Life in the Oceanic Crust: Areas Where Observatories Are Very Useful to Investigate a Particular Scientific Problem and Where They Are Useful

Observatory science is VERY USEFUL to investigate the following:

- The chemical and biological response to episodic volcanic and hydrothermal events;
- The formation of event plumes;
- Subsurface biosphere;
- Marine food webs on the seafloor; and
- The linkages between geological, biological, and chemical processes in ocean crust.

Observatory science is USEFUL to investigate the following:

- Fluid flow on ridge flanks; and
- Simultaneous environmental variability in ridge crest, flanks, and convergent margins.

To address the science where observatories are very useful, development or improvement of the following sensors is needed:

- Chemical, biological, and flow-rate sensors that can operate at high temperatures for long-term deployment without servicing;
- Fluid and biological samplers suitable for long-term deployment;
- AUVs, drifters, and other instruments for use in rapid-event response;
- Downhole sensors for use in boreholes;
- Instruments for acoustic event detection (such as the SOund SUrveillance System [SOSUS]); and
- Advanced capabilities for drilling into ocean crust.

injected into the crust. This research would include the response of seafloor biological communities at convergent margin seepage sites to abrupt changes in fluid and chemical fluxes caused by seismic activity. Similarly, the dynamics of gas hydrate formation and dissociation, especially in response to perturbations produced by tectonic cycles or global warming, is a problem of current interest that could be addressed by observatory science.

FUTURE DIRECTIONS AND MAJOR SCIENTIFIC PROBLEMS

Four different oceanic environments are important for research on fluids and life in the ocean crust: ridge crests, ridge flanks, convergent margins, and coastal areas on passive margins. The ridge-crest environment can be further subdivided into sedimented versus non-sedimented ridges, magma-rich versus

BOX 2-3
OBSERVATORY SCENARIO: FLUID DETECTION, OBSERVATION, AND RESPONSE TO A SEAFLOOR VOLCANIC ERUPTION

Eruptive events on the seafloor have the ability to release great volumes of hydrothermal fluid that affect the chemistry and biology of overlying water and generate a unique type of hydrothermal plume called an event plume (Figure 2-3). Although it is not known how event plumes are formed, it is clear that they are produced by a sudden catastrophic release of large quantities of hot water. Eruptions also extrude lava on the seafloor, creating new habitats for endemic vent faunas and increasing production and export of deep-living microbial populations. Initial changes in water-column and seafloor properties after an eruptive event are very difficult to study using expeditionary approaches, although recent event detection and response efforts at Juan de Fuca Ridge have made good progress toward this end. A seafloor observatory system near a volcanically active site would provide an important platform for characterizing the early stages of an event while also monitoring longer-term changes. For instance, once an eruptive event has been detected (acoustically or otherwise), sensors in the water column and near the seafloor could increase data collection and transfer rates. An Autonomous Underwater Vehicle (AUV) could then be released to image the seafloor at the eruption site to record geochemical and biological characteristics of the water column and benthos. Drifters could also be released from the observatory node to track the event plume as it was advected from the site, and could measure the changes in geochemistry, particulates, and organisms as the plume evolved. On longer timescales, a Remotely Operated Vehicle (ROV) could be mobilized to initiate experiments and collect samples at the eruption site.

magma-starved ridges, and ridges recently perturbed by magma input versus ridges in a mature or quiescent state. Similarly, the subduction zone environment can be further subdivided into convergent margins with sediment accretion versus sediment subduction, fast versus slow subduction rates, and sediment type (terrigenous versus marine). The ridge-flank environment can be categorized according to the nature and thickness of sediment cover, and association with fast or slow spreading ridge. Within each of these environ-

BOX 2-4
NEW MILLENNIUM OBSERVATORY

Objective: The objective of the New Millennium Observatory (NeMO) is to monitor and sample geophysical, geochemical, and microbial variability on an active segment of a mid-ocean ridge system for determination of the relationships between subseafloor magma movement; faulting; and changes in the biologic, chemical, and physical properties of the subsurface biosphere (Figure 2-3). Instrumentation deployed includes acoustic range meters (extensometers), high- and low-temperature probes, osmosamplers, time-lapse cameras, seafloor pressure gauges, microbial traps, and seafloor navigation transponders.

Location: Axial seamount (summit caldera at 1,520 m water depth) on the Juan de Fuca Ridge, approximately 250 mi off the coast of Oregon and Washington

Established: Autumn 1997

FIGURE 2-3 Axial Volcano, Juan de Fuca Ridge—the first place where new tube worms were found on the site of the 1998 Axial Volcano seafloor eruption. It is near the edge of the lava flow (note the new lava in the background). Photo taken in 1999. SOURCE: National Oceanic and Atmospheric Administration Vents Program.

ments, it is critical to determine the nature of and the linkages between tectonic, thermal, chemical, and biological processes at a variety of temporal and spatial scales to achieve an in-depth scientific understanding of the processes that are occurring.

The following are some of the important scientific goals for future research that can be addressed with seafloor observatories:

- assess the extent of the sub-seafloor biosphere and determine its biological and chemical character;
- assess the impact of fluid and gas flow and related processes on crustal structure and composition, ocean chemistry, and biological productivity within and above the seafloor;
- determine the fluid flow patterns on ridge crests, ridge flanks, and in convergent margins through space and time;
- directly observe the changes in heat, chemical fluxes, and biological diversity produced by ridge-crest magmatic and tectonic events;
- determine how hydrothermal-event plumes (megaplumes) form and assess their global importance;
- directly observe how biological productivity and diversity change in response to fluctuations in fluid and chemical fluxes at vents and seeps on ridge crests, ridge flanks, and at convergent margins;
- assess the extent to which sediment cover and spreading rate or subduction rate affect fluid chemistry and biological diversity at ridge crests, ridge flanks, and subduction zones and assess the impact of these chemical systems on the overlying ocean;
- quantify the importance of chemosynthetic productivity on the seafloor;
- understand the relations between tectonic and fluid processes in subduction zones;
- determine rates of gas hydrate formation and dissociation in response to perturbations of pressure, temperature, and fluid chemistry and flow rate and determine the influence on ocean chemistry, biology, and climate.

THE ROLE OF SUSTAINED TIME-SERIES OBSERVATIONS

Previous observations of fluids related to the ocean crust have been made mainly by deploying single, short-duration experiments that stored data rather than transmitted information in real time. To make significant advances in this area of research, it is essential to observe co-varying processes by making synoptic[1] measurements over a variety of timescales. The ability to simulta-

[1]Synoptic - large-scale simultaneous collection of measurements or observations.

neously conduct a variety of sustained time-series measurements will greatly aid the understanding of linkages between geological, chemical, physical, and biological processes. Furthermore, real-time availability of data is essential, as it will allow scientists to respond to unusual events or modify experiments if necessary. Seafloor observatories could also provide valuable real-time control of experiments and allow in situ process studies in which intervention or perturbation experiments could be conducted. On a broader scale, a more ambitious goal would be to design an observatory that could provide simultaneous observations on the ridge crest, ridge flank, and convergent margins. This would provide crucial new insights and understanding of the linkages among these different environments.

TECHNICAL REQUIREMENTS

For future studies of fluids and biota in the oceanic crust, access to the subsurface is a critical requirement. One way to sample and observe fluids and biota in the crust is by drilling, and ideally some seafloor observatory experiments would be conducted in conjunction with boreholes drilled into the crust (Box 2-5). Thus, the continuation of the Ocean Drilling Program and the development of new drilling capabilities will be an essential adjunct to seafloor observatory studies.

The ability to respond rapidly to magmatic and tectonic events is another critical need that necessitates relocatable seafloor observatory capabilities that can be rapidly deployed when an event is detected by remote sensing, such as acoustic detection using submarine hydrophones. Another approach might involve the use of AUVs, already in place at docking stations, that could be called into service to conduct water-column surveys and photoreconnaissance of an event locality. Rapid response is also dependent on continued real-time acoustic monitoring, such as that presently provided by the SOund SUrveillance System (SOSUS) array.

Another important goal is to properly configure an observatory or network of observatories to simultaneously monitor changes on the ridge crest, the ridge flank, and at the convergent margin in real time. This would provide direct observations of the linkages between these different environments and would also allow observations of the entire hydrologic cycle associated with seafloor spreading, from the trapping of fluids in the crust on the ridge crest and ridge flank, to the dewatering of crust and gas hydrate formation and dissolution at convergent margins.

The development of new sensors and observational tools for long-range monitoring in a variety of environments is essential for the observatory approach to be effective. There is a critical need for new chemical, flow-rate, and biological sensors, and also for improved camera systems that are immune

BOX 2-5
BOREHOLE CIRCULATION OBVIATION RETROFIT KITS

From 1991 to 1997, the Ocean Drilling Program installed 13 long-term hydrogeological observatories or Circulation Obviation Retrofit Kits (CORKs) to study fluid flow processes in situ in two documented flow regimes: sedimented young oceanic crust and accretionary prisms. These observatories involve sealing a borehole at the throat of the reentry cone with a sensor string suspended in the hole. A long-term data logger is positioned so that it is accessible by a submersible at the seafloor for periodic data transfer and reprogramming.

Long-term records of borehole temperature and pressure indicate that months or even years were required to reestablish fluid conditions exiting in the formation prior to drilling. Borehole temperature and pressure records also captured the signals of hydrologic transients and nearby earthquakes, and allowed the fluid transport and elastic properties of the CORKed formation to be determined at multiple scales using the spectral analysis of phase and attenuation of seafloor tidal-loading signals that propagate into the subsurface.

Some CORKs also incorporate valves on the seafloor to access the sealed hole either for direct hydrological testing of the formation or fluid sampling for geochemical and microbiological analysis. Most of the sites instrumented with CORKs do not produce formation fluids at the valve, and have required deployment of long-term, self-contained, osmotically driven "osmosamplers" on the sensor strings for recovery of in situ fluids. Hole 1026B on the Juan de Fuca Ridge (Plate III) is an example of a site where crustal fluids are actively produced. This hole was drilled into a topographic high with a permeable basement that focuses circulation of subsurface fluids. This site generated considerable excitement as the location of a long-term, microbiological filtering experiment deployed for over two years by a team from the universities of Hawaii and Washington.

The successes of the original CORK hydrogeological observatories made a significant contribution to the in situ monitoring of geological processes. Building on these successes, a new generation of advanced CORKs has been developed, incorporating a capability to separately seal and isolate multiple zones in the sub-seafloor formation, so the observatory configurations correspond much more closely to natural hydrogeological structures. This new generation will also accept an enhanced range of sensors and auxiliary experiments, allowing customized sub-seafloor observatories for studying a variety of fluids-related problems (e.g., sub-seafloor microbiological communities and processes, gas hydrate processes) or seismogenesis and the balance among tectonic and fluid-flow processes at plate boundary faults.

to biofouling. For studies of subsurface fluids and biota, new downhole sensors and experimental packages are needed for borehole deployment.

Biological studies within the crustal and vent environments will necessitate a wide range of experiment types that in turn will require different observatory approaches. To study the biological response to transient events, such as the colonization of newly formed vent fields, relocatable observatories that can be rapidly deployed are essential. In contrast, at sites exhibiting relatively stable behavior, biological experiments will require a permanent or long-term observatory with high-bandwidth data transmission for high-quality video and other types of monitoring.

DYNAMICS OF OCEANIC LITHOSPHERE AND IMAGING EARTH'S INTERIOR

Increasingly, geoscience research in the oceans is moving beyond the exploration and mapping of the seafloor and is focusing on understanding the dynamics of the solid earth system and the interaction of geological, chemical, and biological processes through time. The expeditionary approach common in the ocean sciences in the past, in which seagoing investigations were conducted in regions for short time periods, is poorly suited for detecting or understanding longer-term change. A continuous measurement presence is required with the ability to react quickly to episodic events, such as earthquakes or volcanic eruptions (Table 2-3). Long-term observations are also needed for making measurements of signals with poor signal-to-noise ratios, or measuring change that is only resolvable over long time periods, such as the motion or internal deformation of lithospheric plates. Geophysical observatories have long been an integral component of earth science research on land; advances in technology and our understanding of the oceans now make it feasible to establish long-term observatories on the seafloor (Boxes 2-6 and 2-7).

MAJOR SCIENTIFIC PROBLEMS AND ROLE OF SEAFLOOR OBSERVATORIES

There are many areas of earth science research that would be advanced through the collection of sustained time-series observations on the seafloor. These include:

Global earth structure and core-mantle dynamics—Fundamental scientific questions currently exist concerning the dynamics of Earth's mantle and core. Until only a few years ago it was generally accepted that subducting slabs penetrated no deeper than the 670 km discontinuity. New results suggest that some slabs descend into the lower mantle, possibly to the core-mantle boundary. This has reopened fundamental questions about the scales of con-

TABLE 2-3 Dynamics of Oceanic Lithosphere and Imaging the Earth's Interior: Areas Where Observatories Are Very Useful to Investigate a Particular Scientific Problem and Where They Are Useful

Observatory science is VERY USEFUL to investigate the following:

- Global earth structure;
- Seismogenesis at subduction zone megathrusts;
- Ridge-crest processes and creation of oceanic crust; and
- Oceanic volcanism.

Observatory science is USEFUL to investigate the following:

- Upper mantle dynamics;
- Ocean plate kinematics;
- Plate deformation; and
- Geologic hazard mitigation.

To address the science where observatories are very useful, development or improvement of the following sensors is needed:

- Long-range, multi-mission, multi-sensor AUVs (including docking capability);
- Advanced ROVs with improved manipulator and cable-laying capability;
- Borehole and buriable broadband seafloor seismometers;
- Seafloor geodetic sensors (strain, tilt, gravity);
- Borehole fluid pressure/composition and samplers; and
- In situ biological and chemical sensors.

vection in Earth's mantle (layered versus whole mantle) and the existence and origin of distinct mantle geochemical reservoirs. These topics will be major focuses of future research.

The nature and origin of hotspots and their interaction with the lithosphere are other important questions in mantle dynamics. New paleomagnetic data are inconsistent with the idea of "fixed" hot spots and geochemical data appear to be compatible with a variety of different origins for hot spots, including the 670 km discontinuity, a boundary between different geochemical reservoirs in the lower mantle, or the core-mantle boundary (CMB). The CMB is compositionally and perhaps dynamically the most dramatic boundary within the earth and it has been proposed as the origin of hot spots and the graveyard of subduction zones. Future studies will be aimed at determining the role of the CMB in the larger-scale dynamics of Earth's mantle.

The earth's inner core, which comprises less than 1 percent of Earth's volume, is the last frontier of solid-earth geophysics. Data that sample this region must travel through the remaining 99 percent of Earth's volume; thus, all of the errors introduced by correcting for the "known 3-D structure of the mantle" project themselves onto the inner core. There are indications that the properties of this region are more complex than previously thought, and as our knowledge of mantle heterogeneity improves so will our resolution of the structure of Earth's inner core.

Four centuries after the demonstration by Gilbert that Earth's magnetic field is largely of internal origin, our understanding of the dynamo process responsible for generating the field remains incomplete. The abundance of data from direct observation of the field over historical time periods, and in particular over recent periods from satellites and permanent land observatories, enables the field to be mapped with far greater resolution than could possibly be realized with paleomagnetic observations. There is a growing recognition from these data that it is the behavior of the field on very short timescales—from decadal to annual periods—that lies at the heart of some of the most important problems in geomagnetism.

Improved spatial sampling provided by long-term seismic and geomagnetic observations in the ocean promises great gains in understanding the geodynamics of Earth's interior and the origin of Earth's magnetic field. Large gaps exist in the global network of seismic and geomagnetic stations that cannot be filled with island stations, particularly in the remote eastern Pacific and Southern oceans. Various studies and workshop reports published over the past 15 years (e.g., Purdy and Dziewonski, 1988; OSN, 1995) have shown that the establishment of ~20 high-quality, broadband (0.003-5 Hz) permanent seismic stations in the oceans would provide much improved tomographic imaging of lower mantle structure (especially in the southern hemisphere) and core-mantle boundary, and would also illuminate the role of subducting slabs and plumes in deep mantle circulation (Plate II). This OSN, first proposed over a decade ago, is envisioned as part of a larger International Ocean Network (ION) (Montagner and Lancelot, 1995) that would also include ~8 seafloor magnetic observatory sites identified by a task group of the U.S. Geodynamics Committee as necessary for an improved global characterization of the short-term behavior of Earth's magnetic field.

Seismogenesis: Subduction zone megathrusts and continental deformation—Plate tectonic theory provides a quantitative framework within which lithospheric deformation and faulting can be understood as well as a first-order explanation for the global distribution of seismicity and earthquakes. However, investigations are increasingly focusing on the deformation process itself and the still poorly understood interplay among tectonic stress, rock rheology, fluid distribution, and faulting. Critical questions include: (1) How do large faults move at unexpectedly low stress levels? (2) Why do some faults lock, eventually leading to large earthquakes, while others slip nearly aseismically? (3) What is the origin of low-angle detachment faults? (4) How is regional stress localized

BOX 2-6
THE HAWAII-2 OBSERVATORY

Objective: A permanent, deep-ocean, scientific research facility, the Hawaii-2 Observatory (H2O), was installed on the retired HAW-2 cable (Figure 2-4) in September 1998. The observatory consists of a seafloor submarine cable termination and junction box in 5,000 m of water located halfway between Hawaii and California (Figure 2-5). The infrastructure was installed from a large research vessel using the remotely operated vehicle (ROV), *Jason*, and standard over-the-side gear. The junction box provides two-way digital communication at variable data rates of up to 115 kb/s using the RS-422 protocol and a total of 400 W of power for both junction box systems and user equipment. Instruments may be connected to the junction box at eight wet-mateable connectors using ROVs. The H2O junction box is a "smart" design that incorporates redundancy to protect against failure and with full control of instrument functionality from shore. Initial instrumentation at the observatory site includes broadband seismometer and hydrophone packages. Further information may be found at http://www.whoi.edu/science/GG/DSO/H2O/ (H20, 2000).

Location: Halfway between Hawaii and California (about 28° N, 143° W) in 5 km of water on a relatively featureless part of the seafloor between the Murray and Molokai fracture zones.

Established: September 1998

FIGURE 2-4 The Hawaii-2 (H2O) junction box on the seafloor as photographed by Jason. This junction box is designed to be removed for servicing and reinstalled by a research vessel and a remotely operated vehicle (ROV). SOURCE: Alan Chave, Woods Hole Oceanographic Institution.

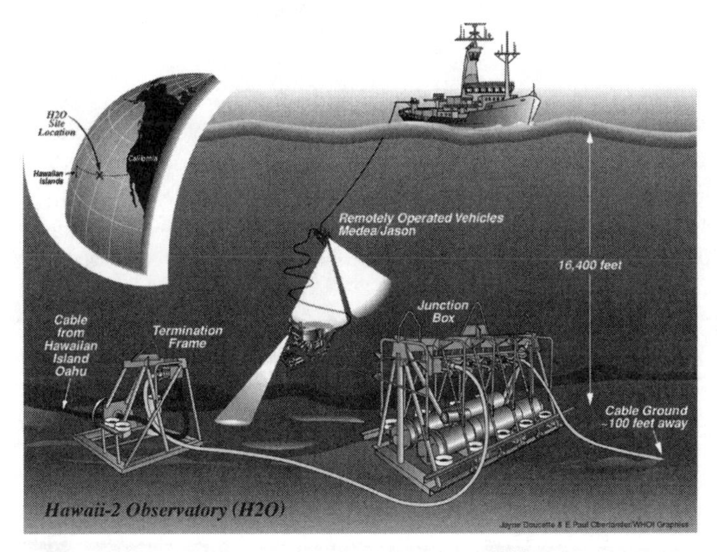

FIGURE 2-5 Schematic drawing of the Hawaii-2 (H2O) observatory and its site. SOURCE: Alan Chave, Woods Hole Oceanographic Institution.

BOX 2-7
THE HAWAI'I UNDERSEA GEO-OBSERVATORY

Objective: The Hawai'i Undersea Geo-Observatory (HUGO) is an observatory installed on the summit of the Loihi seamount, the youngest volcano of the Hawaiian chain. This observatory was designed to provide electrical power, command capability, and real-time data transfer for experiments on Loihi. HUGO supplies up to 5 kW of power and nearly unlimited data bandwidth for experimenters using an electro-optical cable. HUGO experiments, supplied by the science community, are installed by submersible or ROV connection to the junction box and multiplexing nodes (Figure 2-6). Possible experiments include seismometers, hydrophones, and pressure sensors, as well as systems to monitor biological activity, chemical variations, hydrothermal venting, and motion of the unstable flank of Kilauea volcano (Plate IV). The observatory junction box is located at a depth of 1,200 m and has excellent acoustic visibility of approximately 1/4 of the world's oceans. Data can be viewed and heard at: http://www.soest.hawaii.edu/HUGO/hugo.html (HUGO, 1998).

Location: The Hawai'i Undersea Geo-Observatory junction box is located at a water depth of 1,200 m, approximately 30 mi southeast of the island of Hawai'i. The system is operated by the University of Hawaii's School of Ocean & Earth Science & Technology.

Established: HUGO was established in October 1997. The system operated until April 28, 1998, when an electrical fault in the cable shut it down. The system was again operated for a short period in October 1998, when a battery package was plugged into the junction box. Currently, operators are looking at options for replacing the cable and making improvements to the system.

and magnified to initiate faults? (5) Why do "slow earthquakes" appear to occur only along seafloor transform faults?

Subduction zone megathrusts produce the largest and potentially most destructive earthquakes and tsunamis on Earth (Figure 2-7). Despite the obvious societal impacts of these great earthquakes, little is known about the seismogenic zone that produces them. Understanding the origin of major earthquakes at subduction zones, such as those off the coasts of Japan, Central

BOX 2-7 CONTINUED

FIGURE 2-6 The starboard side of the HUGO junction box on Loihi seamount off Hawaii. Experiments are installed and repaired using the Pisces V submersible. SOURCE: Fred Duennebier, University of Hawaii.

America, and Cascadia, is the key focus of the international SEIsmogenic Zone Experiments (SEIZE) initiative. SEIZE experiments are aimed at determining the linkages between large-scale plate motions, strain accumulation, fault evolution, and fluid flow. The scientific strategy employed by SEIZE involves a combination of geophysical imaging, drilling, and long-term monitoring over an earthquake cycle (a few years to several decades) at a few representative subduction zones. The necessary long-term measurements include strain,

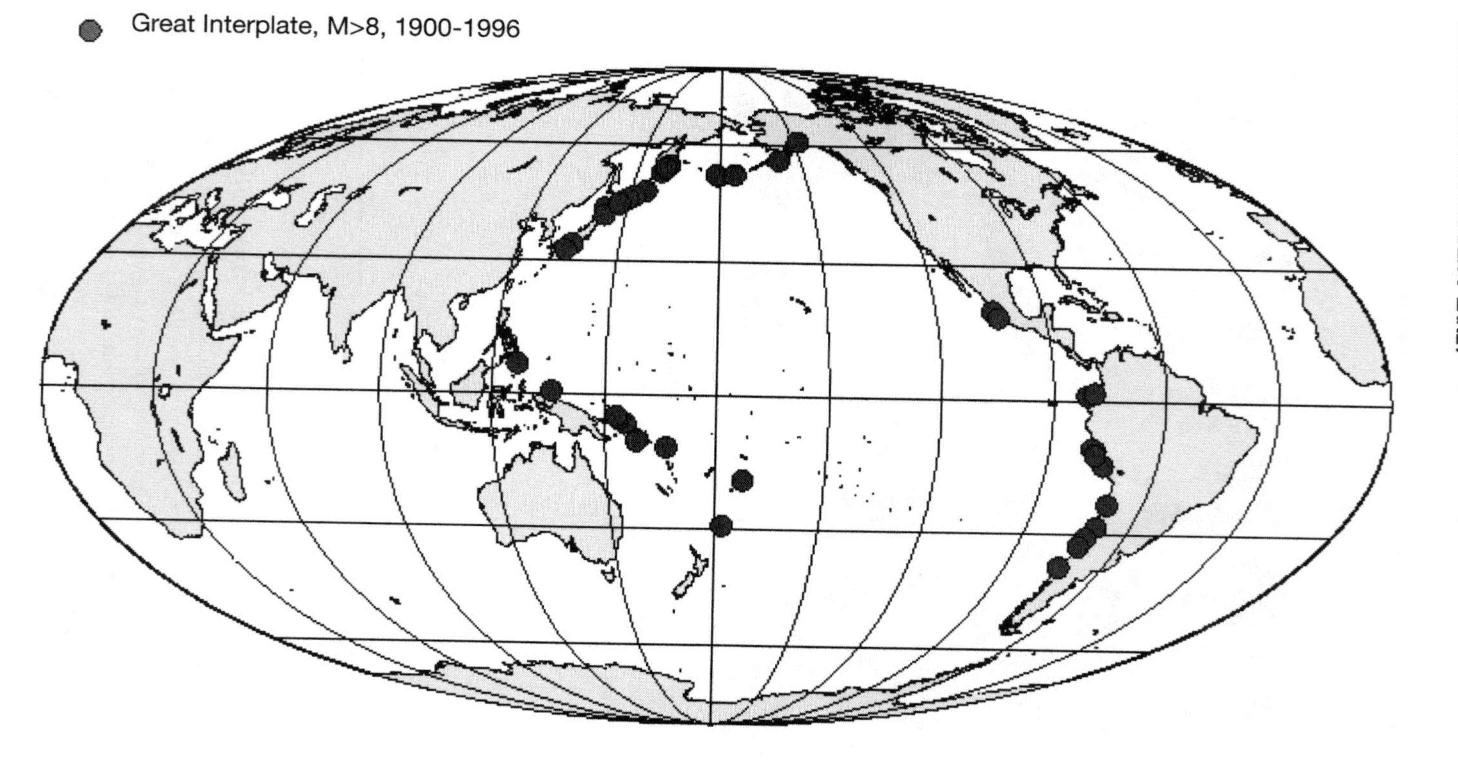

FIGURE 2-7 Global distribution of great interplate earthquakes M>8 (Moore and Moore, 1998).

seismicity, and fluid flow at sites on and beneath the seafloor (in boreholes), as well as on land. These observations, in concert with laboratory experiments on the behavior of material in the seismogenic zone and theoretical modeling, offer the potential for major advances in our understanding of earthquake processes.

Major earthquakes are also common where continental crust is deforming, such as in the Basin and Range province in the western United States and along the San Andreas fault system in California. This area will be a major focus of a new program in the earth sciences community known as Earth Scope, and its subprograms the USArray and the Plate Boundary Observatory. A seafloor observatory program would complement these land-based studies by providing the capability to monitor deformation and faulting seaward of major active fault systems, like the San Andreas, in tectonic settings where fault systems may have less complex structures than on the continent.

Ridge-crest processes and oceanic volcanoes—Volcanism at mid-ocean ridges, in intraplate settings, and along convergent margins controls the flux of heat, mass, and volatiles from Earth's interior. The circulation of hydrothermal fluids transfers heat from the crust to the overlying ocean, extracts or deposits metal sulfides, and modifies ocean chemistry. As discussed in the section titled "Fluids and Life in the Oceanic Crust" earlier in this chapter, these same fluids support a variety of life-forms, from thermophilic bacteria in the subsurface to macro-organisms on the seafloor around vents. Hydrothermal processes affect the chemical, thermal, and biological balance of oceanic environments, and also provide the best insights into primitive earth systems that harbored and conceivably initiated Earth's first life-forms. As a result of successful interdisciplinary programs, such as Ridge InterDisciplinary Global Experiments (RIDGE), and its international cousin, InterRIDGE, the importance of understanding the links among geological, physical, chemical, and biological processes at submarine volcanic systems is well established. However, the specific nature of these linkages and their variation in response to transient events, such as episodic dike intrusions and volcanic eruptions, is still poorly understood and will be a major focus of future studies.

Current ship-based studies allow only periodic visits (at intervals of months to years) to study volcanically active areas, and the deployment of low-(battery) powered, autonomous instruments offer the ability to make continuous measurements for periods of up to only a year or so. Response to remotely detected seismic and volcanic events must be mounted from shore, and can take weeks or months to reach the area of interest. Thus, critical hydrothermal and biogeochemical process that occur at and immediately after the time of an eruption have never been observed.

By installing long-term observatory nodes at 3 to 6 sites along the global mid-ocean ridge system and at a few oceanic volcanoes in other (off-axis) settings, the linkages among these geological, physical, chemical, and biological processes and their response to transient volcanic and tectonic events can be studied. Observatory nodes in active volcanic and hydrothermal areas could provide the power, control, and data bandwidth for interactive arrays of geophysical, chemical, and biological sensors installed on the seafloor or in boreholes. By supporting water-column observations from fixed moorings and from AUVs, the heat and chemical fluxes from chronic or episodic hydrothermal venting can be measured. The observatory node could also power remotely controlled seafloor rovers and AUVs to survey and sample more distant sites. In some cases, available power may permit in situ chemical and biological analyses, as well as the storage or freezing of critical samples for later recovery and shore-based analysis.

Oceanic plate kinematics, plate deformation, and faulting—Our picture of plate tectonics is based largely on historical datasets, such as the geomagnetic reversal record averaged over millions of years, geomorphologic estimates of transform-fault azimuths, and present-day earthquake slip vectors. These data have provided input to a plate tectonic model in which a dozen presumably rigid plates are each assigned a velocity vector. While successful overall, this model lacks detail because, essentially, all of our planet's plate boundaries are underwater. With a few notable exceptions (e.g., the San Andreas fault system), there are few constraints on contemporary crustal motions near plate boundaries.

A number of attempts are underway to begin monitoring motions along plate boundaries using various means, such as tilt, strain, and absolute gravity. All programs are aimed at sampling a single site and, in most cases, at widely spaced intervals in time. Seafloor observatories provide an opportunity to advance the embryonic science of monitoring geodetic motions on the seafloor. It will finally be possible to continuously observe motion near plate boundaries. The key questions seafloor geodetic observatories could address are as follows: (1) How do velocity vectors vary within an oceanic plate? (2) How variable is the deformation rate in the neo-volcanic zone extending from ridge axes down the flanks to the presumed rigid-plate edge? (3) Do spreading centers spread continuously or episodically? (4) In a subduction zone, where does the velocity change from continuous motion to a stick-slip regime? (5) How does deformation of the ocean basins affect mean sealevel? Seafloor geodetic studies will complement major new programs on the continents, including EarthScope and its subprograms USArray and the Plate Boundary Observatory.

Geological hazard mitigation—As the human population continues to grow, the potential social and economic dislocation from natural hazards, such as earthquakes, volcanoes, submarine landslides, and tsunamis, has increased. Major earthquakes in Japan, Taiwan, Turkey, and southern California over the past decade have cost many thousands of lives and resulted in severe economic impacts (estimated at more than $1 trillion for the 1995 Kobe earthquake alone). Recent volcanic eruptions in Indonesia, Nicaragua, Papua New Guinea, and Ecuador have dislocated local populations, damaged infrastructure, and caused significant economic damage. These impacts are especially detrimental to developing nations, such as these, that are not easily able to rebuild infrastructure or mitigate economic effects.

As many of the most seismogenic areas, and some of the world's most active volcanoes, occur along the margins of continents, seafloor observatories will play a key role in global geohazard assessment and monitoring. Networks of hydroacoustic monitors, such as SOSUS, are needed to detect earthquakes, identify submarine volcanic eruptions, locate major submarine landslides, and provide early warning for tsunamis. Regional observatories, for example, those offshore of California and the Pacific Northwest, will be important because of the major earthquake risk in these areas. Relocatable observatories will be useful for detailed studies of active regions. In order to determine anomalous activity that may presage a seismic or volcanic event, decades of data should be collected continuously. A wide variety of sensors will be required to assess geological hazards, including seismometers; acoustic, tilt, and pressure sensors; and fluid monitoring and sampling devices.

TECHNICAL REQUIREMENTS

The diverse solid-earth applications for seafloor observatories result in a range of technical requirements that argue for a flexible observatory infrastructure using both cabled-based and moored-buoy systems.

The technical requirements for the global OSN sites are comparatively simple: a broadband seismometer with a sensor mounted in a borehole or buried in sediment, and a seafloor magnetometer. Power requirements (<10 W on the seafloor) and data telemetry needs (a few 10s of Mb/day) are modest. Continuous real-time data telemetry will be useful, although dial-up capabilities to retrieve data from specific time intervals of interest would be sufficient in many cases, reducing telemetry needs to a few Mb/day or less. OSN sites will offer many opportunities for ancillary measurements for oceanographic (Plate II), geochemical, and geodetic studies that in some cases (e.g., acoustic thermometry) could significantly increase the power and telemetry requirements. Because of the remote locations proposed for OSN sites (typically 1,500-2,000 km from land), moored-buoy installations are likely to be the favored approach in most cases, but in some instances cabled installations may be possible (e.g., H2O [Box 2-6]). It is expected that the OSN sites will need to be serviced at least annually. Given their remote locations (many are in the Southern Ocean), servicing of these stations will be logistically challenging and costly.

A seafloor geodetic experiment would require 3 to 4 battery-operated seafloor transponders, a three-antenna global positioning system (GPS), an acoustic transmitter, and a sonar processor on the buoy (Figure 2-8). Power requirements for the GPS systems would be approximately 10 to 20 W and the acoustic system would likely operate continuously for 48 hours every 1 to 2 weeks. Other geodetic sensors, such as seafloor strainmeters, tiltmeters, and absolute gravimeters are under development, and would be essential components of a seafloor geodetic experiment.

Observatories on ridge crests, at subduction-zone megathrusts, and on mid-plate volcanoes will need to be designed to support a wide range of sensor types. The resulting power and bandwidth requirements could be substantial. For many physical processes of interest, sensors already exist. However, development of chemical and biological sensors is needed to characterize hydrothermal fluids and measure in situ variations in biomass, metabolic activity, and species composition. For certain applications, photography and real-time video will be required. In addition, remotely controlled seafloor rovers and AUVs will be needed to collect data and samples in a broader area around each

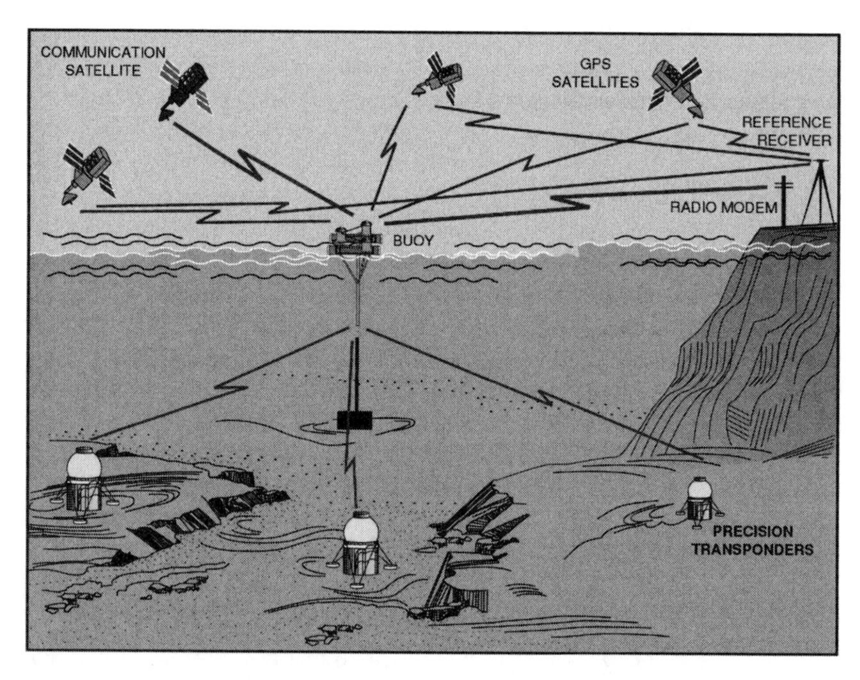

FIGURE 2-8 Schematic drawing of a seafloor geodetic experiment showing the buoy, seafloor transponders, global positioning system, and acoustic transmitter. SOURCE: John Orcutt, Scripps Institution of Oceanography.

node. The ability to drill boreholes for the installation of sub-seafloor instruments will also be an essential requirement.

The power and telemetry needs for ridge-crest, subduction-zone megathrust, and mid-plate volcano observatories will be relatively high (100s of W to several kW of power at each node; 100s of Mb/day to Gb/day of telemetry bandwidth). These requirements will favor cabled observatories for many sites but, for remote locations (e.g., East Pacific Rise or parts of the Mid-Atlantic Ridge), large moored-buoy systems may be more cost-effective. A relocatable and rapidly deployable moored-buoy observatory system will also be needed to study transient natural events (volcanic eruptions, earthquake swarms) and to conduct in situ perturbation experiments. These systems will have comparatively modest power and bandwidth requirements.

COASTAL OCEAN PROCESSES

The coastal ocean includes estuaries, the surf zone, the Laurentian Great Lakes, the continental shelf, and the continental slope. This definition incorporates a great deal of environmental diversity. For example, estuaries have distinctly different environmental characteristics in terms of depth, processes, and accessibility as compared to the continental slope. Furthermore, coastal regions are distinct from the deeper, open ocean for a number of reasons: (1) the coastal ocean is where terrestrial influences encounter the broader ocean; (2) the coastal zone is the area of the ocean most strongly affected by anthropogenic impacts; (3) the coastal ocean displays a strong geographic diversity, depending on forcing agents and topographical setting; (4) the coastal ocean, being generally shallow, tends to be strongly influenced by energy and material fluxes through the surface and bottom; (5) the coastal zone is the most biologically productive part of the ocean and hence is the most heavily fished; (6) physical, chemical, geological, and biological patterns in the coastal ocean tend to be anisotropic,[2] intermittent (both in space and time), and rapidly changing relative to the deeper ocean. These factors make the coastal ocean a challenging environment for scientific study.

FUTURE DIRECTIONS AND MAJOR SCIENTIFIC PROBLEMS

Perceptions of what are the important scientific problems in coastal oceanography are possibly as diverse as the number of investigators. Discussions at the symposium showed, however, that common themes exist and, despite the diversity of goals, there is a striking commonality in terms of the sorts of long-term measurements that might be valuable. Some major problems in coastal ocean science are listed below.

[2]Anisotropy - having properties that differ according to the direction of measurement.

BOX 2-8
OBSERVATORY SCENARIO: DETECTION, OBSERVATION, AND RESPONSE TO AN ALGAL BLOOM

Algal blooms are episodic, not presently predictable, and difficult to study using traditional expeditionary approaches. Blooms of some species have serious impacts on marine ecosystems and human health, and are the focus of general environmental concern (http://www.redtide.whoi.edu/hab/nationplan/nationplan.html [EcoHAB, 1994]). Large algal blooms that occur offshore provide a pulse of organic matter to deep environments, affecting population dynamics and seasonality in bottom-dwelling organisms. A seafloor observatory positioned in a location where blooms occur could contribute substantially to our understanding of processes initiating algal blooms, and the effect of these blooms on marine communities and food webs. Such an observatory might include an array of nodes with sensors measuring physical, chemical, and biological characteristics at the surface, throughout the water column, and on the seafloor. These measurements would likely include temperature, salinity, nutrients, water clarity, and fluorescence. Once a bloom is detected, sensors would increase data collection and communication rates; an AUV would be sent out to survey water-column and seafloor effects; and drifters would be released to track the horizontal advection and dispersion of the bloom. In the near term, a ship might be deployed to take measurements and samples in the bloom-affected water mass; in subsequent years, drifter technology may improve so that all measurements are made by sensors on the drifters. If the bloom is deemed toxic, monitoring of shellfish, fish, and marine mammals would be initiated and pertinent information would be transferred to resource managers and the public.

- Coastal ecosystems are extremely productive, but not very well understood in terms of how their structures and functions respond to variations in environmental conditions. For example, we need to learn more about the causes and predictability of harmful algal blooms (Box 2-8). Part of this learning process would require the collection of long time-series datasets of physical, nutrient, and algal conditions in the coastal ocean that could then be used to enhance our level of understanding, as well as be used for actual day-to-day prediction.

- Global change is of general interest to society. In the coastal ocean, there are concerns both about how such change might affect littoral areas (e.g., how a change of wind patterns affect might coastal productivity), and about how the resulting variations within the coastal zone might affect the rest of the ocean through such processes as water mass formation at high latitudes and coastal influences on global biogeochemical cycles. Separating natural changes from anthropogenic change is essential.
- A number of coastal management issues will require long, high-resolution time series of coastal ocean processes. Fisheries applications are an obvious example, but there are numerous other concerns, such as coastal eutrophication, impacts of transportation, and the effects of nearshore minerals exploitation.

A common thread in all these concerns is the need to quantify and model the fluxes of water, materials, and energy through boundaries in the coastal environment, be they lateral (such as river inflows or transport of oceanic waters onshore) or vertical (through the sea surface or the bottom).

THE ROLE AND DESIGN OF SUSTAINED TIME-SERIES OBSERVATIONS

Oceanographic understanding often results from diagnosing how changes in a natural system occur; a more static approach of simply mapping out the present state of a feature tends to be less productive. The need to understand oceanographic change then implies a need to document environmental changes and their potential forcing agents. Because modes of oceanographic variability can change with time over weeks to years, there is a great need to collect long time series in order to document important variations in critical parameters.

Sampling considerations dictate some stringent experimental requirements. If a strongly nonlinear process, such as sediment transport, is of interest, then episodic (rare but highly energetic) events can be expected to dominate the records. These events, such as the passage of a hurricane, will be short-lived and are likely to be unpredictable over the timescales needed to emplace oceanographic measurements. Thus, to observe the impacts of episodic events, well-sampled, long-term observations are needed (Table 2-4). On the other hand, many ongoing coastal ocean processes tend to be dominated by shorter timescale variability, such as tides or wind events. Thus, to investigate this short timescale variability, sampling on hourly versus longer timescales will be needed, as opposed to the weekly timescales that may be meaningful for open ocean processes.

TABLE 2-4 Coastal Ocean Processes: Areas Where Observatories Are Very Useful to Investigate a Particular Scientific Problem and Where They Are Useful

Observatory science is VERY USEFUL to investigate the following:

- Sediment transport;
- Coastal eutrophication; and
- The impacts of global environmental change on the coastal environment.

Observatory science is USEFUL to investigate the following:

- Fisheries science; and
- The structure and function of coastal ecosystems.

To address the science where observatories are essential, development or improvement of the following sensors is needed:

- Increased duration, speed, and long-term reliability for AUVs;
- In situ nutrient and chemical sensors; and
- In situ zooplankton sensors.

The coastal ocean is often typified by relatively short spatial scales and by strong anisotropy. For example, alongshore currents over the shelf tend to be typified by cross-shelf scales of 10 kms in magnitude, but can have considerably greater alongshore scales. Short spatial scales mean that a single-point measurement may only represent a very small volume of the ocean. Thus, there is a clear need to make sufficient spatial measurements to fully characterize the environment of interest. It is useful to envision a sustained three-dimensional network of measurements that will be able to describe, in detail, a given area and that will allow evaluation of advective transports.

The coastal ocean incorporates a variety of environments. For example, the shelf off the coast of Maine is distinctly different than that off Georgia, and estuarine environments are distinctly different than those of the continental shelf. It appears that the most useful approach to investigate the coastal zone using long time-series measurements is to pick a few representative settings, measure them in detail for a long enough time to gain confidence in having resolved the important variability (10 years would be a credible guess, but this would have to be evaluated as the dataset evolves), and then move the observing capability to another location.

One design for making long-term measurements in the coastal ocean would be to deploy arrays of moorings equipped to measure sediment properties and parameters, such as winds, waves, currents, temperature, salinity, nutrients, optical variables, acoustics, and bottom stress. This list would only represent a starting point, as other measurements might be required for specific scientific needs. The moored array could be L-shaped so as to resolve cross-shelf and alongshore variations. A low-cost model for a moored-buoy observatory has been proposed by Frye et al. (1999). This observatory is portable, easy to deploy and configure, and capable of supporting a variety of instruments, yet it has limited data-transmission capabilities and all instruments must share the same acoustics channel. Another eventual approach would be to deploy a single, well-equipped mooring that is also home to an AUV that can make frequent, rapid surveys of the surrounding region.

Regardless of the design of the measurement network, attention should be paid to coordination with remotely sensed (either from space or shore) measurements and other existing long-term measurements, such as those from tide gauges. Furthermore, real-time communications will be necessary for adaptive sampling within events, and for predictive modeling for coastal management. Improved (or new) sensors are needed for turbulence, dissolved oxygen, plankton, nutrients, and DNA.

TURBULENT MIXING AND BIOPHYSICAL INTERACTION

Turbulent mixing occurs over a broad spectrum of timescales and space scales, strongly affecting the distribution of momentum, heat, chemical compounds, and living organisms in the ocean. Modeling of the physics of turbulent flows is a major limitation in our ability to simulate observed ocean circulation features. Turbulence has widely acknowledged but little studied effects on a range of biological processes, including cycles of surface primary productivity, encounter rates between male and female gametes or predator and prey, aggregation and dispersion of plankton at fronts, and settlement of larvae into the benthos. Turbulent mixing at the sea surface mediates air-sea exchange of biologically reactive compounds, such as dimethyl sulfide and carbon dioxide. Turbulence in the bottom boundary layer plays a role in bentho-pelagic coupling of nutrients, and affects fertilization kinetics, chemical signaling, transport, habitat choice, and genetic exchange in benthic communities. Mesoscale eddies and fronts are now recognized as contributing to patchiness in plankton communities that has long puzzled biologists.

FUTURE DIRECTIONS AND MAJOR SCIENTIFIC PROBLEMS

The general character of turbulent motions is that they are intermittent and random, and thus not predictable in detail. However, prediction of turbulence statistics and its integral impacts on property distributions is possible. Present ocean circulation models very crudely parameterize mixing, with regional and global results being rather sensitive to particular parameterization schemes. Horizontal and vertical turbulent fluxes are treated separately because of the distinctly different importance of vertical stratification and Earth's rotation, and because of the very different turbulence generation mechanisms involved. Because the actual character of turbulent motions in the ocean depends considerably on geographically and temporally variable flow regimes, better parameterizations of the impact of turbulence on oceanic properties are sought first for distinct physical regimes. The grand objective, however, is a parameterization of turbulence statistics as a function of larger-scale, more deterministic flows that can be applied broadly.

The physical oceanography community has identified subregions of general circulation models that are greatly in need of improvement, including deep convection, boundary currents and benthic boundary layers, the dynamics and thermohaline variability of the upper mixed layer, fluxes across the air-sea interface, diapycnal mixing, and topographic effects. Progress in all of these areas is likely as our capacity for modeling smaller scale features increases, and as physical parameterizations are developed (Royer and Young, 1999).

"Vertical" (Diapycnal) Turbulent Fluxes

Major advances in understanding diapycnal turbulent fluxes in the ocean interior have been made recently, and the physical oceanography community is looking forward to a quantum leap in understanding. In *The Future of Physical Oceanography*, this is concisely expressed:

> Past achievements in quantifying small-scale turbulent mixing in the main thermocline, coupled with exciting recent measurements in the deep ocean, suggest that a description and an understanding of the spatial distribution of turbulent mixing in the global ocean is achievable in the next decade. Unraveling the possible connections between the spatial and temporal distribution of mixing, the large-scale meridional over-turning circulation, and climate variability are important aspects of this research. (Royer and Young, 1999)

There is a strong connection between the understanding of coastal ocean processes (see "Coastal Ocean Processes" section earlier in this chapter) and diapycnal mixing:

> Substantial advances have been made in understanding and predicting the peculiar properties of turbulent and rapidly rotating boundary layers over a sloping bottom. Because of the bottom slopes, flow across depth contours implies a vertical density transport which can determine the character, or even existence, of transport in the boundary layer. Flow near the bottom is important because of its role in transporting materials (such as sediments and benthic biota) between shallow and deep water.

> These bottom flows also govern the behavior of stronger alongshore flows in the overlying water column. Thus, an appreciation of the linkage between the bottom boundary layer and the interior, is essential to an understanding of the dynamics of alongshore flows on the continental shelf. (Royer and Young, 1999)

Horizontal Turbulent Fluxes

As with diapycnal turbulent fluxes, there is a clear sense that a quantum improvement in our understanding of horizontal turbulence is within reach:

> Knowledge of the horizontal structure of the ocean on scales between the mesoscale (roughly 50 km) and the microscale (roughly less than 10 m) will be radically advanced and altered. The growing use of towed and autonomous vehicles, in combination with acoustic Doppler current profilers, will revolutionize our view of the ocean by exploring and mapping these almost unvisited scales throughout the global ocean. While this research is driven by interdisciplinary forces (biological processes and variability are active on these relatively small horizontal scales) it is also a new frontier for physical oceanography, and one in which even present technology enables ocean observers to obtain impressive datasets. (Royer and Young, 1999)

THE ROLE OF SUSTAINED TIME-SERIES OBSERVATIONS

The main challenge for the parameterization of turbulent mixing is in obtaining high-quality turbulence statistics and their variation, along with sufficiently dense temporal and spatial information about the oceanic processes that modulate the turbulence. To develop a universal parameterization it is necessary to conduct observational studies in a broad range of environmental conditions (Table 2-5). At present, research cruises of approximately one month's duration are made to a particular region where intensive microstructure profiling is conducted. A vertical profile of turbulence parameters is obtained, along with some sense of its temporal or regional variation. It is usually impossible to separate these variations. In addition, there is often insufficient information concerning the large-scale motions that cause these variations. Because there are relatively few groups that have the technical capability to measure turbulence parameters, the progress of the mixing community has been slow relative to the critical importance of improving numerical models.

While sufficiently long time series are necessary to obtain stable statistical properties of oceanic turbulence, even longer time series are required to observe modulations of these statistics by variations in the large-scale flow environment. However, without the context provided primarily by spatial flux divergences associated with wave propagation and advection, improved mixing

TABLE 2-5 Turbulent Mixing and Biophysical Interaction: Areas Where Observatories Are Very Useful to Investigate a Particular Scientific Problem

Observatory science is VERY USEFUL to accomplish the following:

- Observe and understand processes that modulate vertical turbulence statistics;
- Generalize turbulent flux parameterizations;
- Determine the relationships between temporal and spatial distribution of turbulence in the ocean (assuming AUV capabilities);
- Map subsurface distribution of mesoscale and sub-mesoscale horizontal turbulence (assuming AUV capabilities); and
- Determining the impacts of turbulent mixing on biochemical distributions.

parameterization will be elusive. Because turbulence depends strongly on stratification, vertical resolution of turbulence parameters has the highest priority. Fine sampling in the horizontal will be more valuable than fine temporal sampling for resolving mixing issues.

Repeated, high, vertical-resolution sampling with a coarse, horizontal array of moored conductivity, temperature, and depth (CTD) and velocity profilers is now possible, and such an array can be used to examine mixing issues independent of a seafloor observatory program. To add value to these arrays, the seafloor observatory concept must provide the needed horizontal resolution over much finer scales than presently possible. This increased resolution would provide fully four-dimensional sampling and could be obtained with AUVs. Near real-time telemetry and an active modeling and analysis component could be employed to optimize the sampling strategy. Other methods of obtaining high-resolution spatial information should also be explored, such as horizontal mooring lines and acoustic Doppler technology.

The strategy outlined above could go far in terms of closing budgets or constraining numerical models. (Closure of heat, salt, momentum, and tracer budgets is important because it demonstrates sufficient observational accuracy and adequate resolution of advection and mixing.) A sequence of field programs in different dynamical regimes would provide us with a decadal leap in the understanding of mixing processes by covering the oceanic turbulence parameter space. A seafloor observatory program would be of great benefit to advancing our understanding and parameterization of mixing if it were to enable such a dedicated set of oceanic observations. Suggested dynamical regimes, which should not be considered exhaustive, are discussed in Box 2-9 below.

A similar problem exists for observing and parameterizing the effects of horizontal turbulent motions, in particular for mesoscale eddies, although the technical challenge of observing those scales of turbulence is not nearly so high as for the very small scales of vertical mixing.

BOX 2-9
SURVEYING TURBULENCE REGIMES IN PURSUIT OF UNIVERSAL PARAMETERIZATION

Some of the dynamic regimes that need to be observed to develop more universal turbulence parameterizations include the following:

1) **A smooth bottom dominated by steady geostrophic flows** (e.g., the deep western boundary current below the Gulf Stream).
2) **Rough-bottom topography.** Flow over bottom roughness that extends into the stratified water column can excite internal waves that propagate and break. Special cases are:
   a) **Linear generation models.**
   b) **Nonlinear, finite amplitude bathymetry effects.**
   c) **Bottom boundary condition of downward propagating internal waves.**
   d) **Decay of low-mode tide above smooth bottom boundary**, decoupled from internal tide generation problem.
   e) **Marginal mixing.** Roughness alters the secondary (turbulence-induced) circulations. This appears to be a generic response to mixing about topographically rough environments. Such flows have major implications for biological dispersal.
3) **Dense water formation.** What limits the rate of dense water production and its final density? What role does mixing play in the two-layer outflow? What controls the downstream evolution of the turbulent boundary current on a sloping boundary? What controls vortex shedding and eddy formation, and are they important?
4) **Internal tidal solitons.** Because of their large amplitude and high frequency of occurrence in climatically important regions, the role solitons play in mixing (especially in the near-surface layer) must be determined.
5) **Sill overflows.** Mixing is an important process associated with constrained passages in the abyssal ocean.
6) **Extreme atmospheric forcing of the near-surface boundary layer.**
7) **Double diffusive regime.**
8) **Hydrothermal vent fields.** Vents are a source of turbulence through buoyancy production. Plume dispersal is an important issue for biogeochemistry.

continued

BOX 2-9 CONTINUED

9) **The continental shelf.** How do alongshore flows effect cross-shelf transports?
10) **Vortex shedding.** Flow over and along steep or rough topography can produce vortex shedding, which can efficiently advect and disperse water-mass properties, chemical compounds, and organisms.
11) **Baroclinic instability.** The parameterization of fluxes resulting from this process is very important.
12) **Sediment gravity flows.**

There are important potential synergies that could be achieved by collocating biological and physical studies. These were not addressed by the workshop, but are alluded to in both the *Report of the APROPOS Workshop* (Royer and Young, 1999) and *Report of the OEUVRE Workshop* (Jumars and Hay, 1999). The idea of "imaging" a volume of the ocean to quantify circulation, mixing, biology, and chemistry within a high-resolution cabled observatory has been suggested. The connection of biomass "patchiness" to eddies is an important topic to be pursued.

The turbulent mixing discussion at the workshop concluded that an observational and modeling strategy that uses the seafloor observatory infrastructure could significantly advance our understanding of turbulent mixing processes and our ability to parameterize them in ocean models. This, in turn, would have an important impact on biogeochemical models. The seafloor observatory concept is of limited value to the turbulent mixing community without the ability to implement the strategy of acquiring data in four dimensions while executing dedicated process studies.

TECHNICAL REQUIREMENTS

Platforms for turbulence measurements must be designed to minimize flow distortion and impacts on turbulence structures. Relocatable observatories will be needed to support process studies to survey turbulence parameter space, and to study the influences of turbulent motions on biological processes. To accomplish the required observations within a decade, three observatories must be dedicated to this task, with a deployment of 2 to 3 years in duration at each location. This number of observatories allows for refurbishment and sequential deployments. Each observatory would consist of a coupled array of moorings and a number of AUVs. The moored array would consist of 3 to 5 (subsurface) moored profilers measuring velocity, pressure, temperature, salinity, fluorescence, light transmission, and irradiance; and a surface mooring measuring atmospheric forcing, near-surface currents, and other variables. The mooring could also employ a bottom tripod system, such as the Benthic Acoustic Stress Sensor. The AUVs should measure velocity, conductivity, temperature, depth, and turbulence parameters.

Cabled observatories would be most appropriate for long-term intensive volumetric interdisciplinary studies with evolving capabilities, such as for investigating the interaction of biological processes with mesoscale eddies.

There is a need for turbulence sensors that are compatible with long unattended deployment periods on moorings and AUVs. New sensors must be developed for nutrients (macro and micro) and other chemicals. Sensor development is also required for automated plankton, nekton, and benthic species identification and enumeration by size, physiological state, and other traits of interest. The stability and calibration of all sensors requiring long-term deployment is an important consideration that needs to be addressed. Furthermore, the capability of tracking natural tracers (e.g., atmospheric inputs, seafloor vent effluent) and for releasing and tracking purposeful tracers (SF_6; glass microballs for acoustic tracking) is important for quantifying the integral impacts of mixing.

Technical requirements in common with the ocean climate objectives are given in the section titled "Role of the Ocean in Climate" earlier in this chapter.

ECOSYSTEM DYNAMICS AND BIODIVERSITY

Seafloor observatories are crucial for addressing many of the major scientific problems identified by the NSF Futures report on biological oceanography, *Report of the OEUVRE Workshop* (Jumars and Hay, 1999; Table 2-6). Specific questions benefiting most from the sustained time-series observations are those regarding time-dependent processes and episodically triggered events, and those requiring long-term datasets. These questions fall under broad integrative categories that range from oceanic biology and ecology to biogeochemistry. Other equally valid questions are not well served by the observatory approach, such as those concerned with details of specific biochemical, cellular, or physiological mechanisms occurring within organisms, or behaviors and environmental interactions occurring on the scale of individual organisms. It is also important to note that the observatory approach, while necessary for solving many problems in marine ecology and biological oceanography, is not sufficient alone, and must be used in concert with other approaches (e.g., controlled experimentation, modeling, analysis of the fossil record).

TABLE 2-6 Ecosystem Dynamics and Biodiversity: Areas Where Observatories Are Very Useful to Investigate a Particular Scientific Problem and Where They Are Useful.

Observatory science is VERY USEFUL to accomplish the following:

- Detect and follow episodic ecological events (e.g., plankton blooms, faunal responses to volcanic eruptions or hydrothermal fluid events, faunal responses to detritus-deposition events in deep water, mass spawning events);
- Characterize and understand long-term (annual to decadal) ecological cycles (e.g., predator and prey population dynamics, spread of pathogens);
- Characterize and understand shorter-term (diel, tidal to seasonal) biological cycles (e.g., biogeochemical implications of diel, ontogenetic, and seasonal migrations of populations);
- Detect and monitor ecosystem responses to anthropogenic perturbations (e.g., response of coastal systems to nutrient loading, impacts of large-scale enrichments, influences of climate change on nutrients, trace metals and trace gases, evaluating relationships between environmental forcing functions, and ecosystem state shifts over very long timescales); and
- Forecast population and community changes (e.g., forecasting changes in fisheries stocks and food-web dynamics).

Observatory science is USEFUL to accomplish the following:

- Characterize changes in biodiversity and community structure;
- Determine the spatial scales of the connection between marine populations via dispersal of early life stages (e.g., local population isolation, barriers to dispersal, and linkages in the epidemiology of disease);
- Monitor dynamics of marine food webs (e.g., encounter rates of predators and prey, detection of processes generating large-scale patterns in ecosystems); and
- Characterize gamete mixing, fertilization success, and propagule dispersion.

To address the science where observatories are very useful, development or improvement of the following sensors is needed:

- Long-range AUVs with biosensors and optics;
- Advanced ROVs for episodic sampling, experiment emplacement, and recovery;
- Active tracking sonars whose data can be coupled with satellite imagery;
- Chemical and biological sensors and optics (e.g., spectrophotometers, coulter counters, CHN (carbon, hydrogen, nitrogen) analyzers, video plankton recorders, gene chips);
- Time-sequencing settling plates and particle and organism traps suitable for long-term deployment;
- Both video and still cameras with either sensor or remote control of image collection;
- Active omnidirectional acoustic sonars; and
- In situ sample-processing and sample-collection and preservation capability.

Oceanic ecological observatories will extend into deep water the concept of the U.S. Long-Term Ecological Research Network (LTER), which now includes only a few ocean- and land-margin sites. The mission of the LTER is similar to that proposed for seafloor observatories in that it aims to understand ecological phenomena occurring over long temporal and broad spatial scales and to increase the understanding of major natural and anthropogenic environmental perturbations at selected sites. Just as restricting a seismometer network only to land limits the ability of geophysicists to understand the dynamics of the earth, restricting ecological observatories only to land limits the ability of ecologists to fully understand the dynamics of the biosphere (Box 2-10).

For this discussion, a seafloor observatory is considered in the broadest sense as a system supporting measurements from the seafloor to the ocean surface, including a nested arrangement of instrumentation covering spatial scales of meters to hundreds of kilometers. An observatory might consist of a series of stationary observatory nodes to monitor the seafloor and water column and autonomous underwater vehicles dispatched to provide broader spatial and temporal coverage. This definition does not include Lagrangian drifters or floats per se, but their use will greatly complement an array of fixed observatory sites. In some cases, observatories will be most effective when used with more traditional approaches, such as drifters, ROVs, manned submersibles, and surface ships.

The following are scientific questions well suited for study using seafloor observatories along with a description of the most compatible facility to address the particular scientific problem.

1. How do environmental and biotic factors determine the distributions and activities of key species or communities important to biogeochemical cycles in both space and time?

This question refers to a classical community ecology approach (explaining the distribution and abundance of species) to help understand a systems ecology problem (cycling of mass and energy). An associated question asks what the important interactions are among marine biota, global climate, and biogeochemistry. Important scientific problems related to this general question include responses of coastal margin systems to nutrient loading; biogeochemical implications of daily, seasonal, and life-cycle migrations of populations; food-web analysis; and influences of climate change on nutrients, trace metals, and trace gases. Research under this broad question might deal with the controls on harmful algal blooms, population dynamics of predators, deep-sea community responses to episodic nutrient pulses, or exploitation of fisheries and other resources.

BOX 2-10

LONG-TERM ECOSYSTEM UNDERWATER OBSERVATORY

Objective: The Long-Term Ecosystem Underwater Observatory, 15 m below the surface (LEO-15), provides three subsea nodes for connection of instruments (Figure 2-9). The system provides both power and communications for these instruments via a cable between the subsea nodes at the LEO-15 site and shore facilities at Rutgers University. Each of the identical nodes provides eight standard interfaces for guest instruments as well as a variety of specialized interfaces for other instruments.

Continual measurements made at each node include water temperature, salinity, clarity, wave height, wave period, chlorophyll content, and current speed and direction. An electric/fiber-optic cable that connects to the two permanent subsea nodes at LEO-15 is designed to provide power and two-way, real-time, high-bandwidth communications (including video) to instruments, remote platforms, ROVs, and AUVs. The data received from LEO-15 instruments are used to model and predict currents and summertime upwelling and to aid biologists in the research of benthic communities and phytoplankton ecology.

Location: Immediately offshore of Great Bay near Tuckerton, New Jersey; 15 m of water

Established: August 1996

Long-term, cabled observatories are essential for examining important time-dependent aspects of the question above. This is especially true for those experiments needing instruments with large power requirements. Relocatable observatories will also be a useful approach when addressing ecological questions where shorter-term time series are sufficient (as with the study of hydrothermal vent communities). Global observatory coverage is not

BOX 2-10 CONTINUED

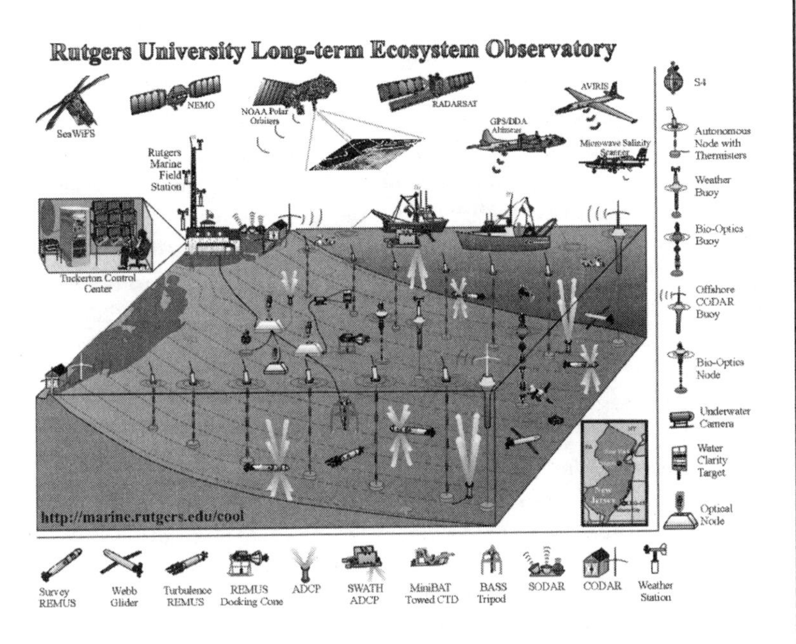

FIGURE 2-9 Diagram of LEO-15 (Long-term Ecosystem Underwater Observatory, 15 meters below the surface). This picture shows the array of underwater instruments currently located and soon to be located at LEO. The fiber-optic cable and nodes were deployed in August 1996. These instruments measure water temperature, salinity, clarity, wave height, wave period, chlorophyll content, and current speeds and directions. The satellite data received in New Brunswick provides information about sea surface temperature, water quality, and phytoplankton content over a huge area (40 deg. latitude x 50 deg. longitude). SOURCE: Scott Glenn, Rutgers University (Rutgers University-COOL, 1999).

necessary for these scientific questions, although a subset of the locations proposed for global climate and geodynamics studies may coincide with sites for these ecological studies.

For certain applications, the sampling interval of data collection may be less than hourly, but data transfer would be required less frequently (e.g., daily or weekly). The main exception to this telemetry requirement would be for applications that include detection and response to episodic events. For these projects, 'smart' instruments should be installed with a capability for quick transmission of data at the onset of an event. As soon as an event or event-precursors are detected, sensors would begin sampling more frequently, two-way communications would be initiated, and associated sample collection or Lagrangian activities (drifters, surface ship) would be mobilized.

Instruments required for this scientific question include stationary sensors recording a variety of physical, chemical, and biological parameters (the latter would include coulter counters, video plankton recorders, etc.); cameras (film, digital, or video); active tracking sonars and other acoustic instrumentation; and AUVs equipped with biosensors and optics. Cameras would most likely be operated in a time-lapse mode, with the exception of those coupled to public and education outreach projects. These would be operated continuously and require real-time data telemetry. Data collected from the sensors listed above should be correlated with satellite data providing optical information and radar imaging for current flow.

It is anticipated that perturbation experiments, such as the controlled release of chemicals or tracers into the water column, or manipulations of seafloor communities using colonization surfaces, faunal clearances, faunal transplants, and predator inclusion and exclusions, will be conducted as part of the observatory activities, and that ecosystem responses will be recorded with sensors and by sampling. Although many studies will continue to require retrieval of water, microbes, plankton, and benthic organisms, this need will decrease as new in situ optical, acoustic, and genetic instruments for detecting and identifying organisms are developed. Samples that are collected could potentially be analyzed in situ using instruments adapted from those used in the laboratory (e.g., spectrophotometers, mass-spectrometers, coulter counters, CHN analyzers, gene chips[3]), or these samples could be preserved using chemical (injection of ethanol, formalin, etc.) or thermal (freezing) means. Alternatively, sampling could be conducted periodically by ROVs during scheduled observatory maintenance visits.

2. What are the functional dynamics of populations and communities?

Specific research topics related to this question deal with the spread of pathogens, the causes and consequences of synchronous spawning, and fertilization success. A longer-term aspect of the question above concerns quantifying the spatial scales over which marine populations are connected via dispersal

[3]Gene chips - This technology promises to monitor the whole genome on a single chip so that researchers can have a better picture of the interactions among thousands of genes simultaneously.

of early life stages. Specifically, research in this area would investigate the influence of local population isolation and barriers to dispersal, and also linkages in the epidemiology of disease. Time-series observations are essential for addressing some aspects of these questions, but they are not sufficient to answer the questions comprehensively. Thus, observatory activities would need to be coordinated with more traditional ship- or submersible-based approaches.

The research questions outlined above would require time-series measurements over a period of one to five years using a flexible (relocatable) observatory system located at strategic, problem-specific locations. Data would be collected at hourly or longer intervals, with the exception of periods following occurrences, such as spawning events, when frequencies of seconds to minutes would be useful to assess such issues as gamete mixing, fertilization success, and propagule dispersion. Data transmission requirements would be modest (on daily or longer frequencies) except during and immediately after events. For fertilization studies, only a few closely spaced nodes would be required, but for studies on the spread of pathogens numerous (on the order of 10) nodes spaced over 10s to 100s of km would be required. Necessary sensors and instruments would include physical, chemical, optical, and biological sensors, such as gene chips, time-sequencing settling plates, and traps. Perturbation experiments would be beneficial to these studies (e.g., induction of spawning through chemical or mechanical shock and controlled release of a non-reactive pathogen mimic), and sampling of organisms for in situ or laboratory analysis (as described above) may be necessary for some projects.

3. What are the dynamics of marine food webs, and how will they respond to environmental perturbations?

Specific scientific problems include encounter rates of predators and prey, long-term variations in population abundances, and detection of processes generating large-scale patterns in ecosystems. Both long-term and relocatable observatories are essential to address this question. Data recording and telemetry frequencies, locations and numbers of nodes and sensors, and methodologies are similar to those described for questions 1 and 2 above.

4. How can population and community changes, such as fluctuations in fisheries stocks and food web dynamics, be accurately forecast? Furthermore, is it possible to evaluate the multiple-scale and pervasive human impacts on the sea, given the confounding effects of weather and climate change?

To address these questions it is necessary to have long-term observations, particularly in relatively unimpacted environments where ecological baselines can be established. It is also important to collect time-series measurements that will illuminate the relationships between environmental forcing and changes in ecosystems over very long timescales. These scientific problems will benefit greatly from coordination within a multidisciplinary approach, including long time-series circulation and climate studies. Cabled observatories would be required for forecasting regional changes, while the coordinated use of data from many widely spaced cabled or moored buoy observatories would be needed to study large-scale physical-chemical-biological interactions.

Long-term studies are likely to require a cabled network with a series of moorings with cameras, active omnidirectional acoustic sonars, and environmental sensors measuring physical, chemical, and biological properties. Because identification, genetic, and chemical analyses of organisms will be a component of the studies, procedures for in situ sample processing and sample collection will be needed. Perturbative and event-response approaches will be less important for these studies, but of possible use in specific projects.

Seafloor Observatory Architecture: Technical Requirements

The principal characteristic of a seafloor observatory, whether it is a moored-buoy or cabled system, is a two-way communication link between instruments and shore. At present, there are two ways to provide this connection: using either a riser or acoustic link from the seafloor to a surface buoy that communicates via satellite or radio to shore, or using a submarine cable linked directly to a shore station.

The following sections outline the characteristics of moored-buoy and cabled observatories, and discuss their current level of technological development. Individual nodes established for various scientific purposes will differ in size, complexity, scientific instrumentation, and technical capabilities in order to balance overall network objectives and cost. This range in the capabilities of individual nodes will permit a measured growth path for the network as a whole. It is envisioned that at the start of an observatory program simpler nodes would be established, with more complex configurations being added when technically feasible. Such an approach will enable the use of knowledge gained from earlier nodes to enhance the later sites. In addition, at the outset of any observatory program, there is likely to be a trade-off between cost and truly global coverage for certain scientific issues. It is important to note that this chapter and the following discuss current and future capabilities and needs related to seafloor observatory infrastructure. The actual designs that would be implemented must be driven by science needs.

MOORED-BUOY OBSERVATORIES

Moored-buoy observatories consist of a surface buoy acting as a central instrument and communications node, with a satellite or direct radio link to shore. This surface buoy is mechanically connected to the seafloor and communicates with instrument packages on the mooring line acoustically or via an electrical or fiber-optic cable. Instruments and observatory devices are either directly connected to the seafloor node of the mooring or communicate via an acoustic communication link. The mooring may or may not supply power to peripheral instruments. This class of observatory system draws heavily on several decades of mooring development and the rapid evolution of satellite communications to provide connectivity from shore to instruments in most ocean regions. In contrast to cabled observatories (discussed below), moored-buoy systems are generally less expensive to install, but the trade-off is a greatly diminished communications bandwidth and reduced power availability.

A wide range of sizes and capabilities can be envisioned for the central mooring. The dominant influences on size are the magnitude of power and communications capabilities needed (Table 3-1). A single mooring design cannot satisfy all science requirements. It is possible, however, to group science requirements into those demanding a large central mooring capable of supplying substantial power (>100 W continuous) or relatively high data rates (>100 Mb/day), and those that can be satisfied with smaller, cheaper moorings, which could be deployed singly or in arrays to provide nested-grid spatial coverage of areas of interest. Smaller moorings are also motivated by a desire to obtain rapid event response (within weeks or months). These relocatable systems might provide a few tens of W of power and support data rates of a few megabytes per day (although not necessarily with continuous data transmission).

Data transfer rate, power consumption, and system stabilization requirements are all interlinked, but it is possible to obtain high data-transfer rates at a relatively modest cost provided compromises are made in power and buoy stability. Some satellite systems, because of the need to use a directional antenna, impose strict stability requirements on the surface buoy (<10 degrees per second in pitch, roll, and yaw). Satellite communications can also consume substantial power. As a result, significant amounts of energy will need to be stored within power sources on the mooring. It is difficult at this stage to make specific recommendations with regard to rapid advancements currently taking place in global digital communications. At present, extremely capable satellite communications systems are under development; however, the success of these ventures is far from certain, as demonstrated by the bankruptcy of the Iridium network.[1]

Data transfer requirements range from less than 1 kb/s to as high as 100 kb/s. In order to prevent disruption to the time-series datasets being collected at a moored-buoy observatory, a backup communications system should be included in the design. Furthermore, the moored system must be capable of storing data onboard if the communications systems were to fail.

[1]Iridium was the first global satellite network for telecommunications. Its 66 crosslinked Low Earth Orbit (LEO) satellites formed a worldwide grid 780 km (485 miles) above Earth.

TABLE 3-1 Power Consumption Versus Bandwidth Requirements for Selected Observatory Instruments

	High Power (> 100 W)	Low Power (< 100 W)
High Bandwidth	Lighting for Photography Imaging/mapping autonomous underwater vehicle (AUV)	Video Photography Hydrophone arrays Seismometer
Low Bandwidth	Water Column mapping AUV Acoustic Thermometry of Ocean Climate (ATOC) source Freezer	Temperature Conductivity Magnetometer Seafloor geodesy Improved Meteorological (IMET) sensors

During symposium discussions, it was assumed that, to keep power requirements low, simpler re-deployable moorings used for rapid event response and habitat studies will not include Acoustic Thermometry of Ocean Climate (ATOC) sound sources, hydrophone arrays, or video. Although video capabilities are desirable for habitat studies, there are practical limitations to providing the necessary communication bandwidth. For re-deployable moorings with AUV capabilities, power requirements are estimated to be 100 W. Without an AUV, power requirements are estimated to be less than 50 W, which is a level achievable using solar panels. Backup battery power will also be required for instrumentation to ensure continued data collection in case of a power failure. Significant advances in power and satellite communications systems will increase the potential capabilities of the low-cost, relocatable moorings.

Moored-buoy systems will need to be engineered for deployment in varied environments, including those at high latitudes. Deployment location will greatly impact mooring design due to variations in sea state, wind velocity, ocean and air temperature, water depth, and satellite communications coverage. Furthermore, consideration must also be given to other factors, such as the suitability of solar panels in high latitudes, potential for vandalism, and visibility of the mooring in shipping lanes. Some scientific requirements may demand an array of deployable moorings spaced between 500 m and 10 km apart, although care must be taken that the minimum spacing between moorings is less than the watch circle of an individual mooring that is itself dependent on water depth.

The establishment of a moored-buoy site will provide capabilities far beyond those of instruments statically connected to the observatory node. For example, the ability to vertically profile physical and chemical parameters in the water column is important for many oceanographic studies. This capability can be provided by placing the vertical profiler on a mooring. While some science problems require profiling the entire water column, many others will focus solely on the near-surface region. Furthermore, many scientific problems outlined in Chapter 2 will require the use of AUVs to sample between moorings and throughout the water column. A moored-buoy system can provide a docking facility for AUVs, where data can be downloaded and batteries recharged.

It is estimated that the average maintenance interval for a moored-buoy observatory will be every 6-12 months. This maintenance will require approximately one week of University-National Oceanographic Laboratory System (UNOLS) ship time in addition to transit time. ROV capabilities are also likely to be needed for observatory servicing. The assessment of specific impacts on the UNOLS fleet arising from observatory maintenance requirements was beyond the capabilities of this committee. The committee believes that data collection capabilities provided by observatories may, however, reduce the demand for ship time and offset the additional time needed for mooring maintenance.

CURRENTLY AVAILABLE TECHNOLOGY

Small Moorings—The oceanographic community has extensive experience with simple moored-buoy systems deployed to acquire long-duration meteorological and oceanographic data (for example, the Woods Hole Oceanographic Institution [WHOI] alone has deployed on the order of 1,000 such systems). These buoy systems fall into two broad categories: surface-following and surface-decoupled systems, depending on whether the surface float closely follows the sea-surface wave motion or whether it is substantially decoupled from this motion. The former systems include disk, boat hull, and toroid-shaped floats; the latter systems include large spar-shaped floats and subsurface moorings. The mooring systems for either category may be single- or multi-leg, but disk buoys have generally been moored with single-point moorings while large spars have often been moored with multi-leg systems. Within the oceanographic community, disk-buoy systems are far more common than spar-buoy systems. The 3 m aluminum disk buoy, developed at WHOI in the early 1980s, has proven rugged and reliable and has been deployed in experiments of up to 2 years at various locations. In the 1990s, the S-tether mooring was developed to provide a platform for the deployment of subsurface instrumentation that needs to be decoupled from the motion of the surface disk buoy. This mooring system is now operational and has been deployed in the Mediterranean and Labrador seas. Moorings with risers capable of supporting power and communication from the seafloor to surface are recent developments, and there have been difficulties making them reliable. Data transfer in a vertical channel from the seafloor to the surface buoy for small moorings is currently provided by acoustic modems that are capable of communication rates of up to 5,000 b/s.

Large Moorings—Large, commercial moored-buoy systems, typically spar buoys, have been in use largely for offshore oil production for some time. The commercial moored-buoy system nearest in capability to high-end requirements identified for an oceanographic observatory network is the OceanNet system recently developed by Maritime Communications Services, a division of Harris Electronics Systems (Figure 3-1). The OceanNet buoy is a 10 m high, large discus buoy that is 5.2 m in diameter and weighs ~110,000 lbs. fully loaded with fuel. This buoy uses a single-point mooring and is equipped with a diesel generator to provide 20 kW peak power and a C-band array of mechanically steered antenna elements capable of telemetry rates of up to 2.2 Mb/s using a commercial INTELSAT satellite. The first OceanNet buoy was deployed in the "Tongue of the Ocean," Bahamas, in 1999 (this buoy sank in a hurricane after 6 weeks on station); a second system is planned for deployment in the Persian Gulf in 2000.

Satellite Communications for Moored-Buoy Systems—At present, the civilian oceanographic community is dependent on commercial satellite communication systems; these systems are described in Chapter 4, in the section titled "Technological Advances in Data Telemetry Technology."

FUTURE DEVELOPMENTS NEEDED

Although much of the technology needed to establish a network of moored observatories already exists, additional developments and enhancements will be needed. For example, communication systems will benefit greatly from planned advances in low Earth orbit satellite systems. Technology enhancements are also expected in the areas of in situ sensor development, data management, and mooring-riser design. To efficiently incorporate technological developments into moored-buoy systems before deployment in remote locations, an easily accessible testing node will be beneficial.

There are several significant development questions that must be addressed before moored-buoy systems can be routinely used as seafloor observatories:

- What are the optimal designs of moored-buoy systems—spar or disk, or some hybrid, such as a spar with a single, S-tether mooring?
- What is the survivability of small spar- or disk-buoy moorings under harsh environmental conditions?

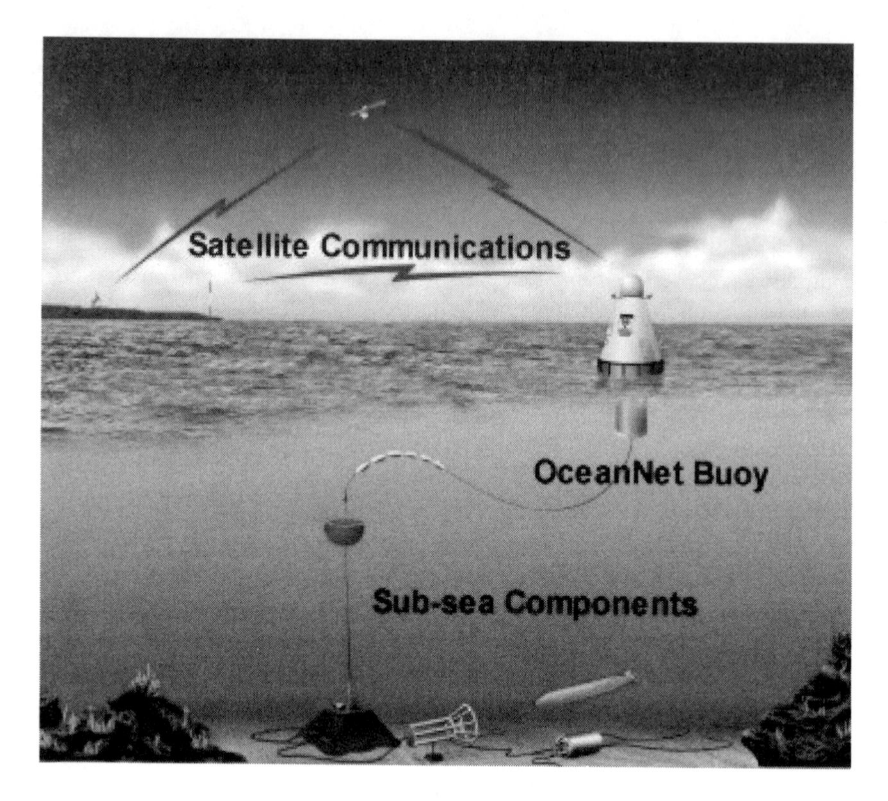

FIGURE 3-1 Diagram of an OceanNet buoy system (MCS, 2000). This system was developed by Maritime Communications Services and consists of deep, shallow, and coastal buoys equipped with satellite communications terminals connected to subsea junction boxes via fiber-optic riser infrastructures that allow direct delivery of data collected. The high performance satellite terminal provides 1 Mb/s throughput and the tuned buoy design allows operation in sea state 6 conditions with a seven-month fuel supply. OceanNet buoys can be deployed in water depths > 3,000 m and can provide global forwarding of data from ocean sensors, custom subsea component interfaces, mission planning and deployment assistance, real-time data collection, autonomous operation, and remote and autonomous command and control.

- How should power and data be delivered between the seafloor and the buoy—fiber-optic or copper-conductor electromechanical cable? How should this cable be protected at critical stress points, for example, near the seafloor and near the surface?
- How far must the junction box be offset from the mooring to allow servicing by a submersible or remotely operated vehicle (ROV)?

- What are the stability and power requirements for satellite data telemetry likely to be in 2-3 years as new satellite communications systems come online?
- What are the expected costs for construction, operation, and maintenance of a moored-buoy, seafloor observatory system? Can UNOLS vessels be used for deployment and maintenance or will specialized vessels be required?
- What are the design, packaging, and transportation arrangements for a rapidly deployed buoy?

CABLED OBSERVATORIES

Cabled seafloor observatories will use undersea telecommunication cables to supply power, communications, and command and control capabilities to scientific monitoring equipment at nodes along the cabled system. Each node could support a range of devices that may include an AUV docking station.

The major components of a cabled observatory will be

- shore stations containing high-power and -voltage, direct-current (DC) generation, network management, and science-experiment management equipment;
- undersea cables containing optical fibers and a power conductor that interconnect shore stations and undersea nodes;
- undersea observatory nodes containing power conditioning, network management, science-experiment management, and standardized science-experiment interfacing equipment;
- network and science-experiment command and control and communications systems;
- specific science-experiment sensors and AUVs.

The requirements of such an observatory include the following:

- *Power*—The power required for a cabled observatory network is greatly dependent on the number of high-energy usage devices, such as lamps, pumps, freezers, and scanners, and is anticipated to range between 2-20 kW per node.
- *Data Transfer Rate*—The communications requirement for a cabled network depends primarily on the use of video transmission and is anticipated to be, at most, less than 1 Gb/s per node.
- *Ships and ROVs*—A commercial cable-laying ship and associated equipment will be needed to install most of the cable, to bury the necessary sections, and to repair cable failures. For operation and most mainte-

nance purposes, one to two dedicated research vessels or workboats with ROVs will be required to support an observatory consisting of some two dozen nodes.

CURRENTLY AVAILABLE TECHNOLOGY

Cable—The current generation of commercial optical undersea cables can satisfy all anticipated observatory data requirements. These cables contain up to eight or a few dozen optical-fiber pairs for long (greater than approximately 300 km) and short systems, respectively, and can provide data transfer rates on the order of 500 Gb/s per fiber pair. Early-generation, commercial cable systems that may soon be retired have a total cable capacity of approximately 500 Mb/s to 2 Gb/s. Such cables can meet many observatory data needs, but it is not clear that they are suitably located or that power capabilities will be sufficient.

Connectors—Undersea mateable connectors suitable for providing electrical connections between the nodes and specific science-experiment equipment are currently available.

Installation and Maintenance Technology—Currently, ships and ROVs that are suitable for installation and maintenance of cabled observatories exist within industry and UNOLS. Although it is likely that UNOLS ship capabilities would be adequate for many maintenance tasks, the additional burden on the already stressed ROV fleet will need to be addressed.

FUTURE DEVELOPMENTS NEEDED

Physical Design of the Cabled Observatory Nodes—Substantial engineering development will be required for the design and packaging of the power conditioning, network management, and science-experiment management equipment to be placed at the observatory nodes. An important component of this development is the design of the thermal management system required due to the relatively high internal power dissipation. In order to meet the necessary specification for high system-operational time (versus downtime), low repair costs, and overall equipment lifetime, significant trade-offs will have to be considered between the use of commercially available and custom-built equipment.

Sensor, Power, and AUV and ROV Technology—The capabilities needed for sensors, power, and AUV and ROV technology are discussed in Chapter 4.

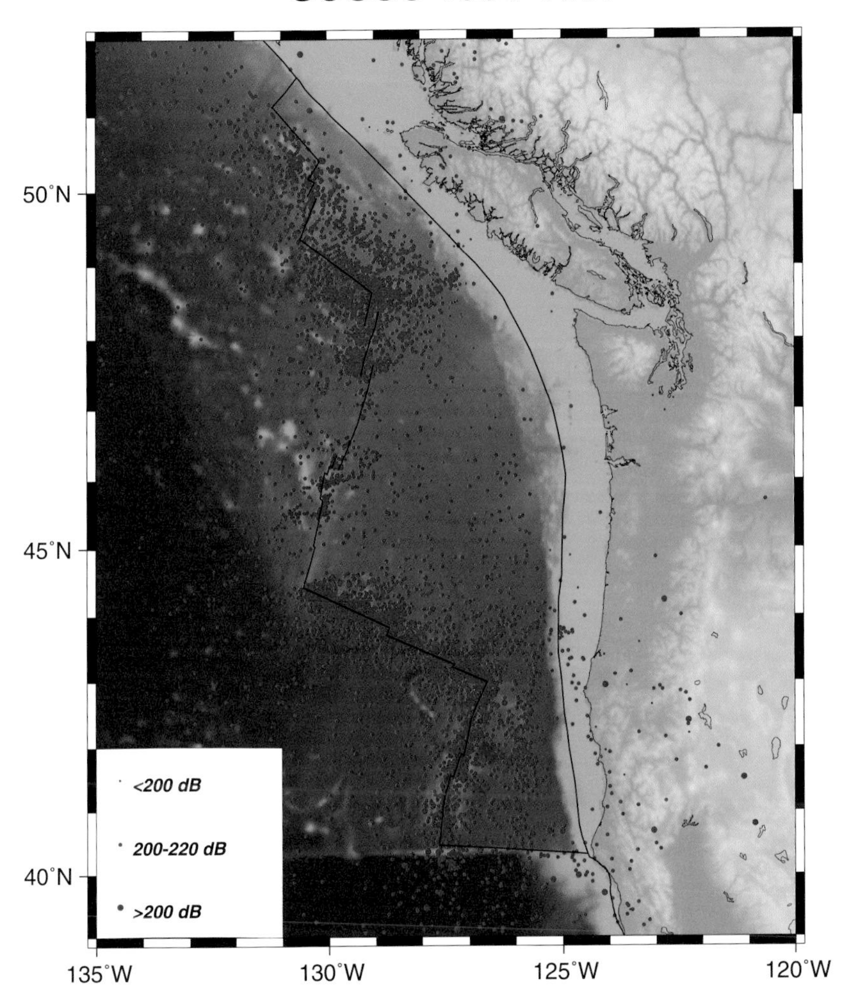

PLATE I The detection of oceanic earthquakes from the SOSUS arrays (3+ arrays from 1991 to 1997). The data set shows high levels of intraplate seismicity within the Gorda and Explorer plates. SOSUS data also reveals significant intraplate seismicity in the Juan de Fuca plate and an alignment of epicenters along the Heck seamount chain. SOURCE: Chris Fox, National Oceanic and Atmospheric Administration's Pacific Marine Environmental Laboratory.

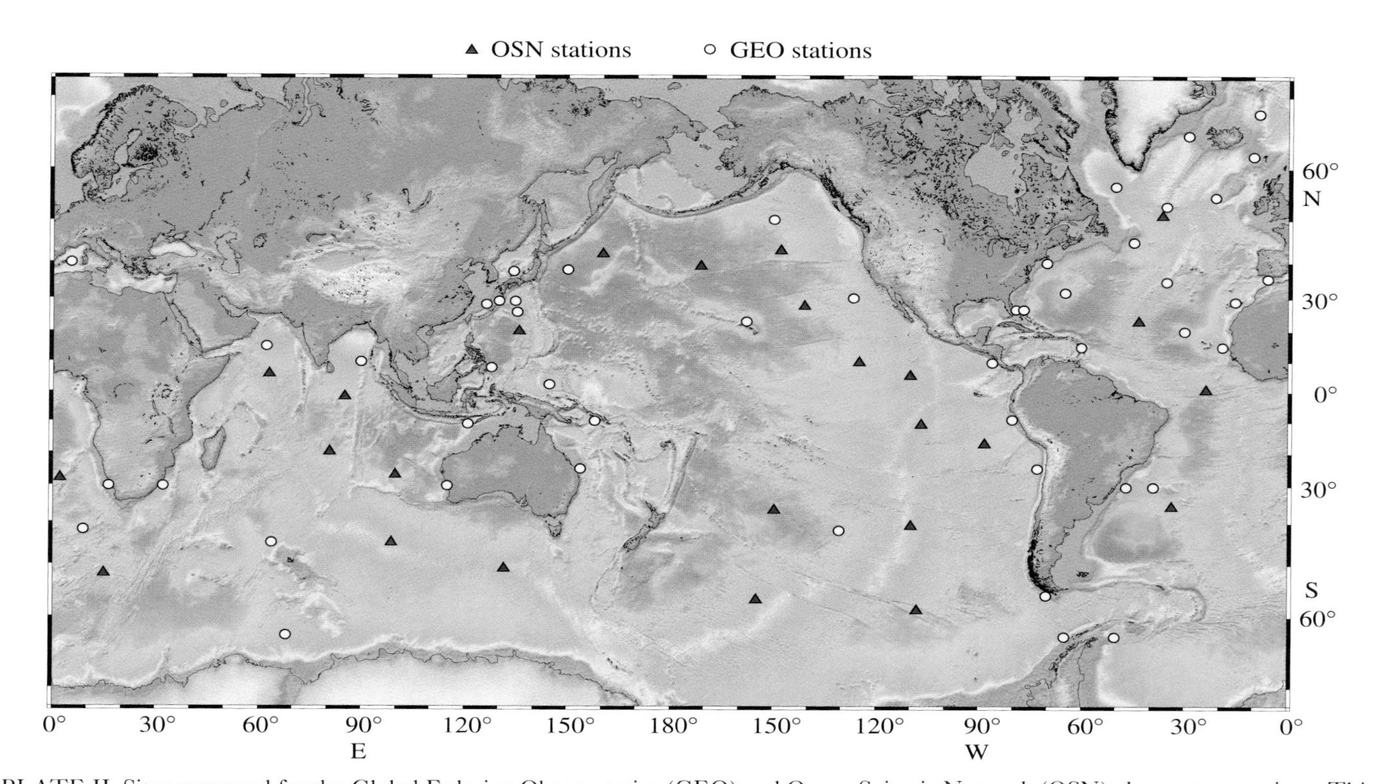

PLATE II Sites proposed for the Global Eulerian Observatories (GEO) and Ocean Seismic Network (OSN) observatory stations. This figure demonstrates that, at many sites, common infrastructure could be used. SOURCE: John Orcutt, Scripps Institution of Oceanography.

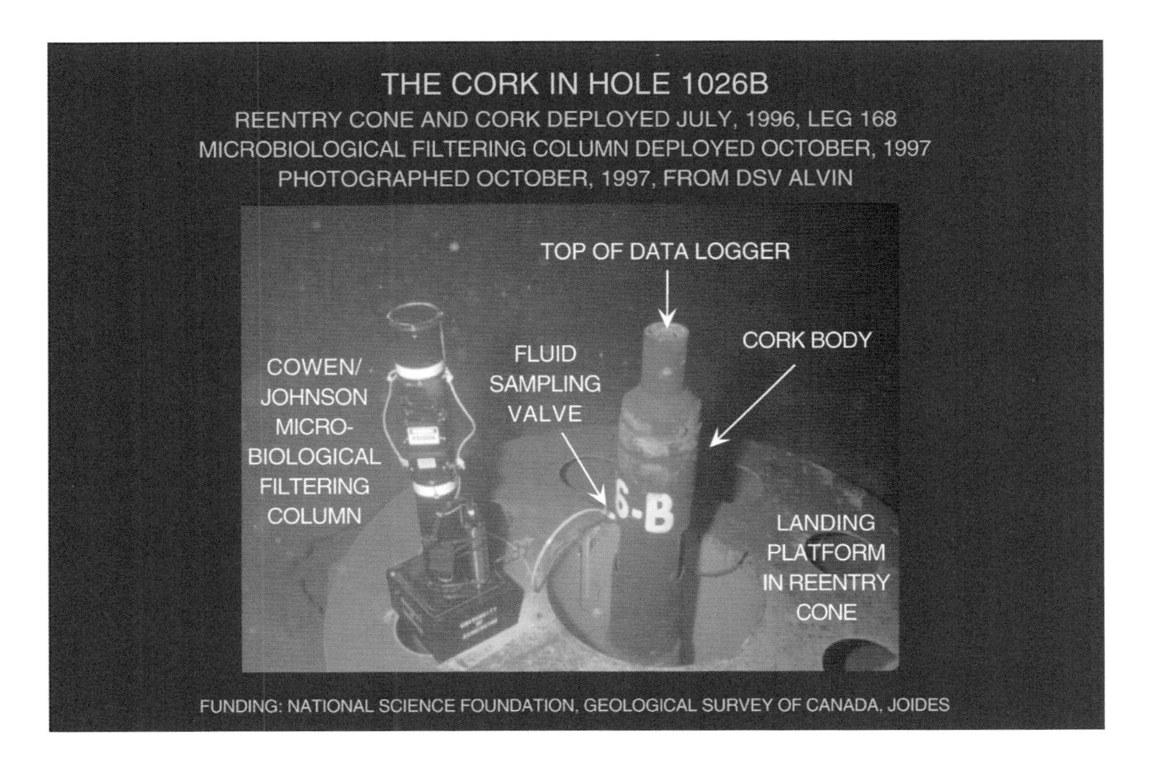

PLATE III A borehole CORK at Ocean Drilling Program (ODP) Hole 1026B. In this photo, the reentry cone, deployed in July 1996 during Leg 168, can be seen. DSV Alvin deployed the microbiological filtering column seen on the left of the photo in October 1997.
SOURCE: Woods Hole Oceanographic Institution and Keir Becker, University of Miami.

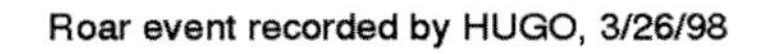

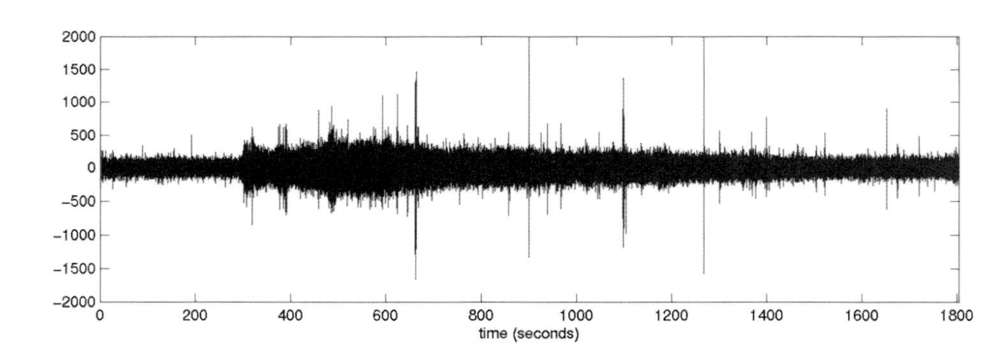

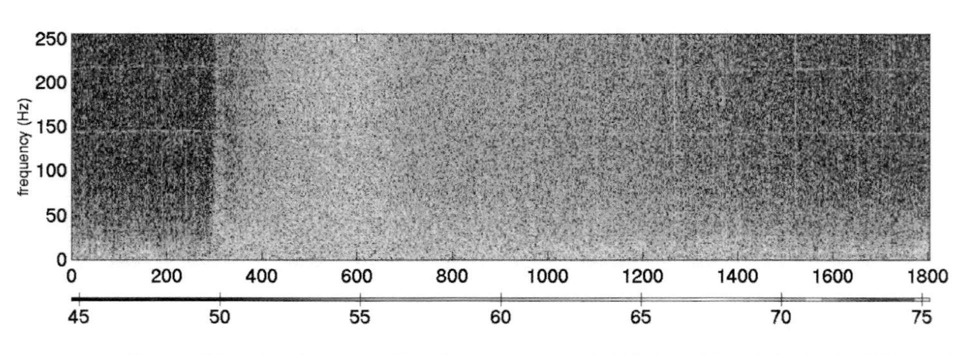

PLATE IV Roar event recorded by the Hawai'i Undersea Geo-Observatory (HUGO) on March 26, 1998. SOURCE: Fred Duennebier, University of Hawaii.

Other Requirements for Establishing a Seafloor Observatory Network

At present, the technology exists to establish a network of both cabled and moored seafloor observatories. This technology includes satellite telemetry, mooring technology, undersea cable technology, and observatory-deployment technology. There will obviously be major improvements in these capabilities and their cost effectiveness, but at present they are sufficiently advanced for general use. As discussed below, many long-term sensors and instruments are also currently available for observatory use (e.g., thermistors, conductivity meters). On the other hand, some sensors are available in basic form, but will need improvements for long-term deployment (e.g., water-particle samplers); whereas others will need major design and development (e.g., many chemical and biological sensors and gene chips). Advances in Autonomous Underwater Vehicles (AUV) technology are also essential for the success of an observatory network. This chapter outlines currently available technology and future developments needed to establish both moored and cabled observatories.

SENSOR TECHNOLOGY

While there are a wide variety of sensors available for undersea research, and there are many key instruments that are being deployed for long time periods (such as seismometers, hydrophone arrays, and current meters), it is clear that development of new sensors will be a critical need in order for seafloor observatories to be fully effective. Sensor technology, particularly in chemistry and biology, is not sufficiently advanced to take optimal advantage of the ocean observatory infrastructure. In addition, enhancements need to be made before existing sensors can be expected to operate unattended for long periods of time in an observatory setting. If an ocean observatory infrastructure is to be established, a substantial parallel investment in sensor technology will be necessary.

CURRENTLY AVAILABLE TECHNOLOGY

Sensors currently available for deployment in a seafloor observatory setting include those for basic physical measurements (temperature, pressure, turbulence, optical clarity, current velocity, wind speed, wave height, seismicity, acoustics, magnetics, gravity), measurements of water properties (conductivity, oxygen, nutrients, dissolved gasses, analysis of suspended particles), and biological measurements (fluorescence, video plankton recorders, hydrophones for ambient noise, acoustic sensors for detection of biota, and gene chips[1]). There are also several types of samplers available for the collection of fluids and biological samples that require shore-based analysis. Many of these sensors and samplers, however, are not presently suitable for long-term deployment for a variety of reasons, such as the need for frequent servicing, instability of chemical reagents, sensitivity to biofouling, or intolerance to high temperatures (for example, there are currently only a handful of properties [temperature, pH, H_2S]) that can be measured in 300°C hydrothermal fluid). Thus, the properties of most high-temperature hydrothermal fluids are presently determined by laboratory analysis of samples after they have cooled. Biofouling is another problem that affects almost all seafloor instrumentation and is especially severe with video cameras, pumps, and other instruments deployed at hydrothermal sites.

In each of the research disciplines, there are certain "basic" sensors that are considered so critical or common that the observatory should supply them as part of the standard infrastructure. At coastal observatories (see Chapter 2, section titled "Coastal Ocean Processes"), these include sensors for basic meteorological conditions (wind speed, air temperature, air pressure); for ocean surface conditions (current velocity, wave height, turbulence); and for water column conditions (temperature, conductivity, nitrate, dissolved gases, pH, gene chip, plankton recorder, water sampler, fluorometer, acoustic fish-tag trackers). For the deep ocean (see Chapter 2, sections titled "Role of the Ocean in Climate," "Turbulent Mixing and Biophysical Interaction," and "Ecosystem Dynamics and Biodiversity"), many critical water-column measurements are similar to those for coastal regions, but they also include sensors for pressure and horizontal electric field, inverted echo sounders, hydrophones, and bioacoustic profilers. In the case of benthic experiments (see Chapter 2, sections titled "Fluids and Life in the Oceanic Crust" and

[1]Although gene chip technology currently exists, it will need significant refinement to be suitable for wide use in oceanographic research.

"Ecosystem Dynamics and Biodiversity"), the most important tool would consist of an AUV equipped with a variety of instruments for imaging the seafloor and collecting samples, including a laser line scanner or sonar system for topography, an imaging scanner for organic matter, in situ sediment sample processing for basic physical parameters, and an acoustic sediment profiler. Other critical benthic instrumentation includes acoustic benthic boundary layer profilers and settlement plates and corers for faunal recruitment studies. At vents, boreholes, and seeps (see Chapter 2, section "Fluids and Life in the Oceanic Crust"), critical instrumentation include those for physical measurements (temperature, fluid flow rate, currents, pressure), for in situ chemical and biological measurements, and for sample collection (osmotic pump samplers, gas-tight samplers, particle samplers). At geophysical seafloor stations (see Chapter 2, section "Dynamics of Oceanic Lithosphere and Imaging Earth's Interior"), basic sensors would include three-component broadband seismometers, hydrophones, magnetometers, absolute pressure transducers, and geodetic sensors for measuring tilt and deformation.

FUTURE DEVELOPMENTS NEEDED

For the full potential of seafloor observatories to be realized, new and innovative seafloor instruments must be developed. Although it is difficult to predict which research areas will give rise to new sensors, there are certain areas where this development is essential. In particular, biological and chemical sensor development lags behind that for physical sensors. Thus, to maximize the potential for biological and chemical research at a seafloor observatory network, sensor development needs to be a high priority.

Sensors that are commonly used only in laboratory settings will require substantial development before these instruments can make accurate measurements while being left unattended for long periods of time. Sensors are also needed that are immune to high temperatures, high pressures, corrosive conditions, biofouling, and other extreme environments found on the seafloor. Furthermore, there is a general need for development of the following: sensors to measure flow rates in high-temperature fluids, advanced sensors deployment for boreholes (downhole logging), instruments for the collection of water samples over extended periods (such as osmosamplers), instruments to collect improved geodetic measurements, and instruments and techniques for the burial and borehole emplacement of broadband seismic sensors.

Considerable development is also needed in the area of ocean acoustics for observatory applications. Hydrophone arrays, such as the SOund SUrveillance System (SOSUS) have been extremely effective at detecting seafloor volcanic and tectonic events and allowing a timely response to these events. Thus, it is recommended that the existing SOSUS arrays be maintained if at all possible and that new relocatable hydrophone systems be developed to augment the fixed SOSUS hydrophones.

An important component of the design phase of an observatory network will be the development of specifications for interfaces between the observatory node and instruments and sensors. These specifications include standardized direct-current (DC) voltages and power control conventions, a layered set of communications protocols using conventional networking standards, and standardized connectors and pin configurations. Standards on allowable environmental effects also need to be established to minimize the impact of an individual instrument or sensor on other observatory experiments. In addition, as part of an observatory network, a certified testing capability will be needed to test instrumentation and identify potential interference problems.

POWER GENERATION TECHNOLOGY

CURRENTLY AVAILABLE TECHNOLOGY

As described previously, the power requirements for a moored-buoy observatory span a range of a few 10s of watts up to several kW. A 50 W output can be achieved with solar or wind power generation, or both, coupled to rechargeable batteries. Power generation above 50 W can be realized using a diesel generator similar to those operating in Coast Guard Large Navigational Buoys (LNB) (Dewey, 1974). The latter approach will obviously entail periodic refueling. Projected power requirements for cabled observatories are anticipated to be 2-20 kW per node. Currently, high-power DC-generation equipment used in commercial cable systems have the capability to provide on the order of 10 kW at approximately 10 kV and 1 A.

FUTURE DEVELOPMENTS NEEDED

For a cabled observatory, significant engineering development will be necessary to provide more than a few kW of power. If the network requires more than approximately 10 kW per shore station, new DC power-generation equipment will be needed. This could be an expensive undertaking because of the limited number of vendors of such equipment and the critical safety concerns. Another consideration is that the power transmission schemes used in the subsea telecommunications industry may not be adequate for a cabled observatory with many nodes. Consequently, power schemes based on existing land-based systems may need to be adopted.

The following developments to seafloor power hardware will be required:

- The current design of specific DC conversion hardware will have to be altered to obtain the required reliability. Engineering this change should be a relatively inexpensive but essential activity.

- To provide a workable thermal environment for the power conditioning, network management and science-experiment management equipment, the physical design of the thermal paths from the electronics to seawater need to be carefully engineered. To meet the necessary thermal specifications, significant trade-offs will have to be made between the use of commercially available equipment and that which is purposely built in order to achieve operational requirements for system operational time (versus downtime), repair cost, and lifetime.
- The observatory physical power path will need to be designed to minimize the probability and effects of corona.[2] The elimination of potential corona generation sites will require careful attention to specifics of the power generation hardware. In addition, the susceptibility of network management and science experiment equipment to corona noise will need to be examined.
- The configuration and hardware for power surge protection need to be designed to provide a fault-tolerant observatory network. It must be assumed that there will be cable and other equipment failures that will produce significant power surges resulting from the interruption of the DC path. If the observatory nodes are not properly protected, there is potential for these surges to seriously damage the node electronics. The development of a surge protection configuraton will require careful computer modeling and simulation of the various fault scenarios. These scenarios could be quite complex for an observatory with many nodes and more than one shore station. Computer simulations will also provide input to the design of the appropriate voltage ramp-up and ramp-down sequences. There is potential for simulation and analysis tools from terrestrial systems to be adopted for these purposes.

TECHNOLOGICAL ADVANCES IN DATA TELEMETRY TECHNOLOGY

CURRENTLY AVAILABLE TECHNOLOGY

Moored-buoy observatories will depend on commercially available satellite communication systems to telemeter data from the buoy back to shore. The satellite communications industry is currently undergoing rapid change as part of the worldwide explosion in wireless communication. The communi-

---

[2] Corona (discharge) - an electrical discharge accompanied by ionization of surrounding atmosphere [syn: corona, corposant, St. Elmo's fire, Saint Elmo's fire, Saint Elmo's light, Saint Ulmo's fire, Saint Ulmo's light, electric glow].

cation rates and tariffs supported by some current and proposed satellite systems are shown in Table 4-1. There are, in addition to global systems (Table 4-1), a number of regional satellite systems that may be capable of providing service to ocean areas, particularly those close to shore. At present, the only global systems in operation that can provide the high bandwidth needed for continuous data transmission are INMARSAT-B and C-Band (Table 4-1). A limitation of these two systems is that they require the gyro-stabilized antenna on the buoy to be pointed with an accuracy of a fraction of a degree. This results in substantial power consumption. Such a system could be mounted on either a disk or spar buoy, but for full-time operation of the satellite telemetry system, the power requirements would necessitate the use of a diesel generator. In addition, tariffs for global coverage are high, making continuous, real-time data telemetry prohibitively expensive. Currently, the only economically viable option for continuous telemetry is to lease time on a C-Band commercial satellite.

It appears likely that, within a few years, there will be operational systems that will be competitive with C-Band and INMARSAT-B in terms of communication rates and possibly superior in terms of power and antenna steering

TABLE 4-1 Specifications for Current and Proposed Global Communications Satellites

SYSTEM	INMARSAT-B[a]	TELEDESIC	GLOBALSTAR	C-BAND
Service types	Voice, fax, data	Voice, fax[b], data[b]	Voice, fax, data	Data
Data Rate	64 kbit/s	64 kbit/s to 2 Mbit/s	9.6 kbit/s	19.2 kbit/s- 2 Mbit/s
Tariff	$7/min $15/Mbyte	Unknown	$1.50-$3.00/min[c] $20-$40/Mbyte	$5K/month/ 128kbit/s[d] $0.015/Mbyte
Service scheduled	In service	Under development	Q4 2000	In service

[a]Tariff would likely be lower for continuous service; the system is presently designed to serve as a dialup resource.
[b]Fax and data service not yet offered.
[c]Service in oceanic areas will be limited, as there must be a gateway within the footprint of a single low Earth orbit satellite.
[d]Obtained by leasing segments from providers, such as INTELSAT, PANAMSAT, and Palapa C1. The segment can be time multiplexed between buoys to reduce overall costs.

TABLE 4-2 Low-Speed Communications Satellites

SYSTEM	INMARSAT-M	ORBCOMM
Service types	Data	Data
Data Rate	2.4 kbit/s	2.4 kbit/s
Tariff	$3/min $167/Mbyte	$10 kbit/s $10K/Mbyte
Service scheduled	In service	In service

requirements (e.g. TELEDESIC). While satellite systems now online or coming online (e.g., GLOBALSTAR) do not have the communication rate capabilities of INMARSAT or INTELSAT (or other dedicated commercial systems), they do have the advantage of requiring only an omnidirectional antenna for data transfer speeds as high as 64 kb/s. The power requirements of these "new-generation" satellite communication systems are projected to be modest (e.g., 14 W transmit; 4.5 W receive), and they will have the capacity to telemeter a substantial subset of observatory data at an acceptable cost.

In addition to higher data rate systems, there are currently lower-speed, low-power satellite systems available for communications and telemetry (Table 4-2). These could be used for applications, such as ARGO, that do not require high telemetry data rates, or as a backup system for a moored buoy.

FUTURE DEVELOPMENTS NEEDED

The satellite communications industry is currently undergoing rapid change, and the systems that will be available in 3 or 5 years are difficult to predict. Market forces outside the academic community will drive the pace of technical development in this industry.

The availability of "new generation" communications systems with low-power, omni-directional antennas and data telemetry rates comparable or superior to current INMARSAT-B or C-Band satellites would make it feasible to design smaller, lighter, lower-cost, moored-buoy systems. It is, however, not certain whether commercial systems will have adequate coverage in areas of scientific interest. If competition with other wireless communication systems reduces tariffs, continuous telemetry at relatively high data-transfer rates may become cost-effective. Given the uncertainty in this rapidly changing industry, the development strategy for moored-buoy observatories should remain flexible.

AUTONOMOUS UNDERWATER VEHICLE TECHNOLOGY FOR SEAFLOOR OBSERVATORIES

AUVs have the potential to undertake a variety of mapping and sampling missions while using fixed observatory installations to recharge batteries, offload data, and receive new instructions (Box 4-1). A principle use for AUVs will be to map seafloor and water-column properties and to document horizontal

BOX 4-1
AUTONOMOUS OCEAN-SAMPLING NETWORKS

Objective: Autonomous Ocean-Sampling Networks (AOSN) are a class of relocatable observatory in which mobile platforms supported by a communications network provide a nested observational capability. This program has developed a significant fraction of the technologies that in the future could support the inclusion of Autonomous Underwater Vehicles (AUVs) in ocean observatories. The AOSN utilize small, high-performance AUVs to provide platforms for a wide range of sensors and sampling systems. These vehicles operate at 3-4 knots for periods on the order of half a day. Docking stations are being developed to provide long-term deployments of these AUVs. Buoyancy-driven vehicles (gliders) operate at speeds less than a knot for a period of several months, carrying a minimal oceanographic sensor suite. Because such vehicles are less expensive, they can be employed in greater numbers to provide a synoptic picture of an ocean region. Development of acoustic communications has been a major objective, to allow real-time control of AUVs and other enabling capabilities, such as adaptive sampling. A multi-institutional effort led by the Massachusetts Institute of Technology has employed a mooring with a docking station and two-way satellite communications capabilities in tests of a relocatable AOSN. Furthermore, the Office of Naval Research (ONR) and Naval Oceanographic Office support for AUV operations at the Long-term Ecosystem Underwater Observatory (LEO-15) (Box 2-10) has demonstrated the utility of AUV operations at a cabled observatory.

Results: Although primarily an engineering-development program, a number of successful AOSN deployments have been demonstrated. Specific elements, such as some acoustic systems and AUV designs, have been licensed to industry and are commercially available.

variability, which is necessary for establishing the context of point-like measurements made from fixed instrumentation. Another use of AUVs will be to extend the spatial observational capability of seafloor observatories. Many oceanographic processes occur episodically, in relatively localized regions. For example, eruptions at spreading ridges may result in the creation of square kilometers of seafloor, but these events may only be indirectly detectable at a mooring tens of kilometers away. AUVs provide a method by which an observatory can deliver instruments to sites of interest, thus greatly expanding the region of coverage.

Scenarios for employing AUVs as elements of seafloor observatories envision small vehicles, weighing at most a few hundred kilograms, with docking capabilities. Docking capabilities are necessitated by the limited endurance of present AUVs, which have typical mission lengths on the order of a day or less. Vehicle endurance is dictated by survey speeds (typically about 5 km/hr) and power consumption by onboard computers and sensors, which can range up to hundreds of watts depending on the payload. The docking capability thus provides the means to extend the AUV presence in the ocean, while at the same time retaining the ability to retrieve data and exert control over AUV missions. Docking also provides a safe parking location for an AUV between activities.

General goals of AUV missions include the following:

Seafloor mapping—AUVs are capable of operating much closer to the seafloor than most other survey systems; therefore, they can collect high-resolution, high-accuracy mapping data, in addition to other geophysical parameters, such as bathymetry and magnetics.

Water-column mapping—AUVs provide the capability to map physical and chemical parameters horizontally by flying a grid at a constant depth, vertically by flying a yo-yo pattern, or in three-dimensions (3-D) by combining vertical and horizontal patterns. Repeated 3-D surveys of an AUV in a given volume enables the production of a 4-D dataset.

Measuring fluxes—Many scientific questions require measurements of the transport of energy, inorganic and organic matter, or dissolved chemical species via a wide range of mechanisms. AUVs provide this capability at specific locations near an observatory node.

Initializing and constraining models—Real-time nowcast and forecast systems for physical parameters (e.g., temperature, salinity, currents) can be useful for a wide range of oceanographic studies. AUVs are capable of obtaining the spatially separated physical measurements needed to initialize and constrain such models.

While AUVs have been employed for oceanographic purposes since the mid-1960s, it is only in the last several years that they have begun to be adopted by the academic oceanographic community. As with Remotely Operated Vehicles (ROVs) in the early 1980s, AUVs are becoming more reliable, and are proving to be useful for a variety of applications. Thus, specific capabilities for AUVs in scientific research are becoming clear. Over the next decade, we can anticipate continuous improvements in endurance, sensing capabilities, navigational infrastructure, communications, and intelligent control of these vehicles.

Existing AUVs exhibit variations in weight of four orders of magnitude, from 10 to 10,000 kg dry weight. However, most U.S. oceanographic efforts are focused on vehicles at the low end of this scale, with most vehicles weighing less than a few hundred kilograms. Although a fuel-cell-powered vehicle with an endurance of up to two weeks is in development, the endurance of a typical AUV is on the order of half a day at speeds of approximately 5 km/hr. A variety of instruments have been used on AUVs, including those to measure conductivity, temperature, and depth (CTD); optical water-quality sensors; turbulence; and current strength and direction (Acoustic Doppler Current Profilers). In addition, still and video cameras, multibeam echo sounders, sidescan sonar, and laser line-scan imagers have also been mounted on AUVs. (NRC, 1996b)

There is an emerging class of AUVs with very long endurance that propel themselves by modulating buoyancy and translating vertical into forward motion through lifting surfaces. These vehicles, commonly referred to as gliders, operate at speeds on the order of 1 km/hr with minimal payloads, to obtain an endurance of many months. The first multi-day glider deployment in the ocean occurred in 1999. Since this deployment, glider capabilities have been progressing rapidly. For the purposes of this report, however, gliders are considered complementary rather than integral to seafloor observatories.

Seafloor observatories will place significant navigation demands on AUVs, but they also could provide an unparalleled navigation infrastructure to support AUV operations. A common navigation technique used in the deep ocean is long-baseline acoustic navigation. Long-baseline navigation employs arrays of bottom-mounted transponders operating at frequencies from 7 to 15 kHz to provide locations with a potential accuracy of several meters. Since transponders are typically placed no further than twice the water depth apart, and much closer for high-performance navigation, it is unusual to provide coverage for areas more than 10 kilometers. The extensive use of such an array, as would be required with repeated AUV operations in a defined area, would likely encourage the deployment of large numbers of transponders for accurate wide-area coverage despite the costs this level of detail would entail. Variations include using lower frequencies to achieve larger separations between transponders, usually at the expense of accuracy and update rate, and using arrays of hydrophones to track acoustic sources as is done at U.S. Navy tracking ranges. The prospect of spatially distributed, electrically connected sites on the seafloor provides opportunities for innovative acoustic navigation techniques that might take advantage of this unusual arrangement.

Another navigation capability for AUV uses inertial and Doppler velocity-log sensors to provide an accurate dead-reckoning capability. Military systems have demonstrated navigation accuracy better than 0.1 percent of distance traveled. The full suite of such sensors and navigation algorithms are expensive and power-consumptive, on the order of hundreds of thousands of dollars and one hundred watts, respectively. However, substantial commercial pressure for this capability promises to deliver more reasonably priced systems with military-level performance and lower power consumption within the time frame of ocean observatories. One problem with this technique is that the AUV must remain sufficiently close to the seafloor for the Doppler velocity log to have bottom lock—a few hundred meters for current systems on the small vehicles discussed here. A correlation velocity log can provide a similar function to the Doppler velocity log at distances much greater from the bottom; however, such systems are not readily available commercially. A final problem with dead-reckoning is that navigation errors are unbounded. This can be remedied by surfacing for a Global Positioning System (GPS) position update; however, for a deep-operating AUV, the transit to and from the surface will be both time-consuming and could introduce navigation errors.

Docking is a capability central to incorporating AUVs into seafloor observatories. This is a particularly complex capability, as it requires several levels of interaction between a vehicle and a docking facility. These include the following:

- *Homing:* The vehicle must be capable of finding its way to a dock.

- *Capture:* The vehicle must be capable of latching to the dock. If it misses, it must be capable of detecting a missed approach to try again.

- *Connection:* A physical connection must be established between the dock and the AUV to provide power and communication links and to physically constrain the vehicle.

- *Data download:* Docking provides a mechanism for downloading data from the vehicle and into the data storage facilities on the dock so that at least some subset of those data can be transferred back to shore. Without such a capability, the value of the data stored on the vehicle will discourage further use of the AUV in case of loss of the vehicle.

- *Battery recharge:* Docking provides an opportunity to recharge the batteries of an AUV, which greatly increases the number of surveys an individual AUV can achieve.

- *Mission upload:* To realize the full flexibility of an AUV, the capability to upload new missions to the vehicle from the dock and, by extension, from shore should be established.

- *Undocking:* Initiating a new mission will involve the release of the vehicle from the docking station.

- *Acoustic communications:* For some applications and classes of observatories, it would be attractive to have a continuous communication link between the vehicle and the observatory.

All the capabilities above have been demonstrated; however, it is likely to take several years of sustained effort to make these capabilities routine. Docking capabilities are currently receiving substantial attention and investment from groups outside the research community, such as the U.S. Navy. This external support should help expedite the development of this critical capability.

Inclusion of one or more AUVs in an observatory can substantially impact the design of the observatory. Not only do AUVs have significant power needs, but the data produced by AUVs can impose substantial demands on observatories relying on satellite communications for data transfer. A rough order-of-magnitude appreciation for these issues can be gained by considering two AUVs types. A water-column AUV might be equipped with sensors consuming little power and producing a low volume of data. An imaging AUV might be equipped with a sonar system that would consume substantial power and produce gigabytes of data.

A water-column AUV might log data at a rate of 10 Mbytes/hr; however, a scientist on shore might be satisfied with a very small subset of this data. For example, a time stamp, position, and three sensor readings every 10 seconds with single-floating-point precision would result in approximately 10 kBytes/hr. The effective average data rate is further throttled by intermittent AUV operations. If one runs the vehicle 1 hour out of 10, then the resulting data stream would be 1 kByte/hr on average. Power consumed by this AUV might conservatively be on the order of 200 W, which at 5 km/hr translates to 40 W/hr per kilometer of distance surveyed. Assuming an operational profile of 1 hour of operation out of 10, the average power required from the dock will be greater than 20 W by a factor relating to the efficiency of power transfer and battery recharging (probably no worse than a factor of 2).

A sonar imaging platform might produce the same amount of data as discussed above, plus an additional GByte/hr for a system such as sidescan sonar. Clearly, this is a high-bandwidth system, even if one succeeds in substantial data compression. Furthermore, the power consumption of such a vehicle could be closer to 300 W, resulting in a correspondingly higher power demand on the dock. For a moored observatory, the data produced by such a vehicle would have to be stored at the docking station (storing it on the vehicle would create the risk of losing all data every time the AUV was operated), creating a substantial mass data-storage demand.

The rapid advance of AUV capabilities is being led by a strong international community of AUV developers. For the most part, oceanographic and military uses of AUVs have been pioneered in the United States, funded primarily by the Office of Naval Research, the Naval Oceanographic Office, and the National Science Foundation. There are several academic research groups in the United States that are capable of fielding AUVs for oceanographic operations, but given the commercial availability and relative affordability of AUVs, the number of institutions with operational oceanographic AUV capabilities is likely to rise dramatically in the next few years.

Recently, the offshore oil and gas industry has demonstrated a growing interest in AUVs for deep-water surveys driven by the combination of the move of offshore drilling to greater water depths and the increasing maturity of AUV capabilities. In the late 1990s, activity such as the Shell Oil development of the Mensa field in the Gulf of Mexico demonstrated that highly profitable oil and gas fields existed in depths in excess of 2,000 meters. These developments and the realization that the deepwater part of the ocean represents the largest unexploited region of the Earth, led to a number of oil companies demanding that their contractors be capable of supporting extraction operations to 3,000 meters. One critical requirement for this capability is the creation of high-resolution maps for planning and installing subsea production facilities. Such surveys are presently accomplished with towed platforms. However, AUVs offer the prospect of achieving both increased economic performance and higher data quality. Motivated by these factors, two marine survey companies have been contracted to construct AUVs for deepwater survey service. Depending on the success of these initial forays into the AUV arena, there is the potential for the multi-billion-dollar offshore oil industry to become significant AUV users.

Historically, the offshore oil and gas industry has not directly invested in AUV development, but this is changing rapidly though the network of contractors who provide equipment and services and who are highly motivated to develop such capabilities. Marine survey companies are driven by the desire to gain a competitive edge in acquiring oil industry contracts, AUV manufacturing companies are motivated by the sales prospects, and sensor and subsystem supplier companies also want a share of the new market. While this has seeded a wide range of technology development, it is important to note that the economic drivers for offshore operations are very different from those of a seafloor observatory network as discussed here. Consequently, the primary effect of the advent of oil and gas industry activity in the AUV area will be the creation of a broader base of expertise and technology rather than providing direct solutions for oceanographic problems.

FUTURE DEVELOPMENTS NEEDED

AUVs today are typically serviced at frequent intervals. To realize their full potential, AUVs must be capable of remaining underwater and operating for extended periods without human servicing. Furthermore, docking remains an experimental capability that must be made reliable and efficient to enable extended, frequent operations of AUVs within a seafloor observatory. While existing navigation and power systems are sufficient to facilitate AUV use at seafloor observatories, improvement in these areas could greatly improve the utility of AUVs.

REMOTELY OPERATED VEHICLE TECHNOLOGY FOR SEAFLOOR OBSERVATORIES

ROVs are likely to play an important role in installing, servicing, and repairing seafloor observatories. Fortunately, ROV technology has been advancing rapidly, both within the oceanographic community and the commercial sector. A wide range of ROV capabilities are available, from systems optimized for high-fidelity control and mapping, to so-called work vehicles capable of delivering hundreds of horsepower of useful work at great depths. While observatory requirements for ROVs are difficult to predict, the current pervasive use of these systems lends a high degree of confidence that the oceanographic and commercial sectors should be capable of supporting seafloor observatory initiatives.

The largest pool of deep-rated ROVs is in industry. Oceaneering alone has an inventory of 76 ROVs with depth ratings of 3,000 m. However, the oceanographic community is routinely using a growing number of deep-rated ROVs, and these systems are increasingly in high demand. Table 4-3 lists systems rated deeper than 1,500 m.

Some common uses for ROV technology include the following:

High-resolution site mapping—High-resolution seafloor maps are important for site preparation prior to the installation of a seafloor observatory. Higher-resolution seafloor maps are obtained by placing mapping sonar systems on platforms flying relatively close to the seafloor. While towed vehicles are routinely used for producing these surveys, the highest quality maps are produced from vehicles capable of flying precision tracks, such as ROVs and AUVs.

Installation of instrumentation—Many types of scientific instrumentation will require installation on or near seafloor observatories. Only a fraction of this instrumentation will exist on or near the central node. Many other instruments will be placed at considerable distances from the node. Because of this, it will be important for an ROV to be capable of spooling out several kilometers of

TABLE 4-3 Deep-rated Remotely Operated Vehicles

Vehicle	Depth Rating	Operating Institution	Builder	Support Vessel
Dolphin 3K	3,000 m	JAMSTEC	JAMSTEC	Natsushima
Hyperdolphin[a]	3,000 m	JAMSTEC	ISE	Kaiyo
Jason	6,000 m	WHOI	WHOI	Atlantis II/ Ship of opportunity
Jason II[b]	6,500 m	WHOI	WHOI	Atlantis II/ Ship of opportunity
Kaiko	10,700 m	JAMSTEC	JAMSTEC	Kairae
ROPOS	5,000 m	Canadian Scientific Submersible Facility	ISE	Ship of opportunity
Ventana	1,830 m	MBARI	ISE	Pt. Lobos
Tiburon	4,000 m	MBARI	MBARI	Western Flyer
Victor	4,000 m	IFREMER	IFREMER	Ship of opportunity

[a]Delivered but not yet operational. Additional vehicle purchased as a spare.
[b]Under development. Projected to be available

fiber-optic cable with low-power conductors to trickle-charge batteries. Scientific packages could be configured as drop-able tool sleds or could be free-dropped from the surface, maneuvered into place by an ROV, and connected to the central node. As proven by the Japanese, it is currently feasible to move large instruments to specific sites after surface deployment using ROV-controlled buoyancy packages. On occasion, seismometers and similar instrumentation may need to be buried or coupled to the seafloor. ROVs with a jetting capability can easily accomplish these tasks.

Servicing installed instrumentation—Even instrumentation positioned deep in the water column can experience biofouling, sediment accumulation, or short-term failures. Standard methods of mounting instrumentation and connector protocols can make instrument replacement routine using a variety of ROVs. There is currently an American Petroleum Institute standard for ROV intervention that should be reviewed for applicability to seafloor observatories. Data downloads are also an important component of ROV servicing that are already being accomplished using inductive modems. Furthermore, short-distance laser data downloads could be used where applicable. It is standard practice in the oilfield to use a hydraulic lance to actuate, latch, and release functions for instrumentation. Seawater hydraulics would provide an environmentally acceptable solution to latching or releasing packages.

Servicing node components—Large nodes should be designed to allow for modular replacement of major components by an ROV. This could be accomplished by unlatching components using seawater hydraulic lances and underwater mateable connector interfaces. This capability will lead to extended service lives for seafloor observatory components.

Burying cable—In certain areas, such as continental shelves, cables will need to be buried to reduce damage by anchors, fishing trawlers, etc.

ROVs are becoming mature enough that, with proper design of observatory hardware, there should be little risk in using ROVs to service observatory systems. The oceanographic community has already demonstrated the use of ROVs in various aspects of observatory work, such as the Woods Hole Oceanographic Institution (WHOI) installation and servicing of Hawaii-2 Observatory (H2O) and Monterey Bay Aquarium Research Institute (MBARI) work in Monterey Bay with the Ventana and Tiburon ROVs. Furthermore, the offshore oil industry is increasingly placing complex systems on the seafloor in deep water that demand ROV serviceability. The commercial undersea cable industry has a set of ROV tools, including plows and tractors, to provide cable burial, maintenance, and repair capabilities. Consequently, there is a growing industrial base on which to build observatory ROV capabilities.

5 Other Significant Issues Relating to Seafloor Observatories

PROGRAM AND PROJECT MANAGEMENT

Given the wide range of disciplines that will be involved in a sustained observational presence on the seafloor and the investment required to establish seafloor observatory facilities, there is an unprecedented need for multidisciplinary coordination of scientific objectives and infrastructure needs. This scientific program and the resulting observatory infrastructure must be driven by high-quality, proposal-driven science. It is also imperative that each observatory site be located, configured, and used for multidisciplinary studies of ocean and earth processes occurring at various time and space scales. These objectives cannot be achieved without effective program and project management.

Management requirements for a seafloor observatory program will change significantly during the evolution from the planning phase into the initial development, design, and implementation phase, and then into the operational phase with subsequent development, design, and implementations. To properly administer a seafloor observatory program, several levels of management will be required, ranging from scientific oversight and systems engineering of distinct observatories, to management of a national observatory infrastructure co-supported by several agencies, to international coordination of global observatory programs. Independent scientific and engineering structures should also be established to determine the scientific direction of the observatory program and to ensure strong project risk management and systems integration.

It is beyond the scope of this report to outline a management structure for a national seafloor observatory initiative. The Committee does recommend that a necessary first step in establishing a seafloor observatory initiative is the development of an implementation program that includes plans for a management structure. During the development of this implementation plan, other relevant agencies should be consulted to ensure cooperation on areas of common interest, such as data management. Interagency discussions will also help identify areas of potential conflict. The Committee suggests the following principles to guide the makeup of a program and project management structure:

- In all phases of a seafloor observatory program, management must be as responsive as possible to the scientific community it serves. As such, the scientific community must be fully involved in the development of an initial program implementation plan and must play a critical oversight role in the establishment and operation of a seafloor observatory network. This role should include the solicitation and coordination of scientific proposals that will drive the development and expansion of the seafloor observatory network.
- Likewise, management must be responsive to the funding agencies supporting seafloor observatories. If co-funding from multiple agencies is involved, as appears will be likely, then interagency agreements must be established as part of the development of a program implementation plan.

The initial development, design, and implementation phase for a seafloor observatories network should include engineering development, prototyping, manufacture, and installation of basic observatory infrastructure. Management of this phase should focus on the following:

- identification and prioritization of specific observatory projects and the development of timetables for their implementation;
- regular review of engineering development, systems integration, prototype efforts, and engineering risks by independent engineers and potential scientific users;
- the completion of previously identified milestones prior to proceeding further with program implementation;
- standardizing of design features common to both cabled and moored-buoy observatories wherever possible, as well as common development of observatory components;
- the development of international interface standards to enable deployment of scientific instrumentation on any type of seafloor observatory;
- the development of a certified testing capability and standardized operational parameters for scientific equipment to minimize impacts on other observatory experiments;

- early identification of common instrumentation needs to advance the development of scientific measurement capabilities in parallel with infrastructure development;
- the establishment of successful demonstration projects before committing to major installation efforts;
- ensuring sufficient flexibility in the program management structure to evolve as development proceeds and scientific goals and opportunities are further elucidated.

A possible program model for the initial development, design, and implementation phase might involve the establishment of a scientific oversight committee and program office with distinct project efforts for observatory programs prioritized for initial development and deployment. Once the initial seafloor infrastructure is successfully installed, program management will shift toward more routine operational, administrative, and management functions, which should be guided by the following principles:

- As seafloor observatories will require tremendous investments in infrastructure, a project management plan must be established to both oversee infrastructure operation and maintenance and to facilitate scientific proposals by independent investigators wishing to use that infrastructure.
- One of the highest priorities of project management should be to ensure fair and equitable access to observatory infrastructure by all funded investigators.
- Project management should encourage the use of seafloor observatory infrastructure by all potential scientific investigators by providing free access to information concerning observatory infrastructure (such as protocols for attaching instruments to observatory nodes) and to all data collected by core observatory instrumentation.
- Given the potential for synoptic measurements with seafloor observatories, it will be particularly important to facilitate truly interdisciplinary projects.
- Access by educators, students, policymakers, and the general public via the Internet must be encouraged.

Currently, other large, coordinated scientific programs in earth, ocean, and planetary sciences (e.g., University-National Oceanographic Laboratory System [UNOLS] ship operations, Incorporated Research Institutions for Seismology [IRIS], Joint Oceanographic Institutions for Deep Earth Sampling [JOIDES] and the international Ocean Drilling Program [ODP], and National Aeronautic and Space Administration [NASA] mission planning) share similar objectives to those listed above. Thus, the management systems of these programs provide useful models for a multi-agency, international seafloor observatory program, and many of the most successful features of these models should be considered, such as the following:

- establishment of a prime management organization by an incorporated consortium of research institutions, universities, agencies, and partnerships that is under contract to the National Science Foundation (NSF) and cooperating agencies (e.g., JOIDES, IRIS, and CORE);
- establishment of a governing body for the incorporated consortium that is composed of experts in science, technology, engineering, and management pertinent to seafloor observatory science (e.g., IRIS);
- oversight of scientific planning by committees of experts appointed by the governing consortium following recommendations from member institutions (e.g., JOIDES);
- awarding of subcontracts to separate operators for distinct observatory projects to encourage competition and the establishment of high-quality efficient services (e.g., UNOLS ship operations);
- awarding of subcontracts to industry when it is best qualified to deliver cost-effective services, such as for engineering development, cable maintenance, large buoy construction, etc. (e.g., the ODP drill ship model);
- allocation of significant resources to data archiving and dissemination from the outset of the program (e.g., IRIS);
- allocation of significant resources for establishment of a high-quality public outreach program (e.g., NASA);
- dedication of significant resources to engineering development of sensors and measurement systems, either in academia or industry as appropriate (e.g., as with the development of ODP third-party downhole logging tools);
- the guarantee of major, sustained agency funding commitments for observatory science based on independent peer review of the observatory management structure and the scientific justifications and outcomes of the observatory program (e.g., ODP).

During the initial stages of this study, the committee was charged by NSF to provide an "order of magnitude" estimate of the costs needed for the establishment and initial phase of a seafloor observatory program. To this end, there was much discussion in the technical breakout sessions at the symposium to attempt to gauge these costs. The outcome of these discussions was that the initial cost of this program could approach several hundred million dollars, with annual operation and maintenance costs approaching tens of millions of dollars annually.

INFORMATION MANAGEMENT

As with program and project management, it is also beyond the scope of this report to detail specific information management protocols. Despite this, from discussions at the symposium and Committee deliberations, certain guiding principles can be suggested.

The three components of a visionary data and information management system that could be incorporated as part of a seafloor observatory program include Internet services, data processing and management, and long-term stewardship. Internet services should include a quick-look, versatile graphics interface with discovery, browse, subsetting, ordering and delivery, interactive visualization, data mining, analysis, and data-driven numerical modeling capabilities. Data processing and management functions should include such items as ingest processing, quality control, reprocessing, product generation, automated catalogs and inventories, browse image generation, and system navigation.

The data collected at seafloor observatories should be centrally managed, with a short time interval between receiving the data and making it available to end users. All data should be preprocessed and permanently archived, and should be non-proprietary unless collected from experimental sensors or individual investigator-initiated experiments. In the latter case, data will remain proprietary for an agreed duration but not to exceed two years, as is consistent with present federal ocean data policy. Raw data and metadata from such experiments should be archived at the earliest possible time to prevent loss. Core financial support should be made available to principal investigators for production of data products suitable for public distribution. Furthermore, support should be given to facilitate educational outreach using seafloor observatory infrastructure, where schools could have direct involvement in the planning and execution of the experiment and receive real-time data. It will also be important to ensure that the data collected from the research-based observatories proposed here can be easily integrated with the data management strategies proposed for the Global Ocean Observing System (GOOS).

Objectives for seafloor observatory information management:

- encourage public involvement by providing as much near real-time information as possible;
- support educational uses for seafloor observatory information;
- provide both the scientific community and the public an efficient method of obtaining information from seafloor observatories;
- support the flow of data from seafloor observatories to data processing nodes;
- assist researchers in the integration and management of their datasets

by providing support for the transition of experimental information into synthesized data products;
- provide the up-to-date status of observing elements and maintenance schedules.

Principles of seafloor observatory information management:

- Routine (facility-generated) information products *and* data should be made available without restriction as quickly as possible.
- Data from experimental sensors or individual investigator-initiated experiments should be made publicly available after quality control procedures have been applied, but within one to two years of retrieval.
- Data summaries (graphics, photographs, etc.) and metadata should be made available in near real-time whenever possible.
- To the extent possible, existing data management facilities, such as the National Oceanographic Data Center (NODC) should be used and enhanced as necessary to meet the needs of seafloor observatories.

A possible model for a "Data, Information, and Knowledge System" for a seafloor observatory program is provided by the Space Physics and Aeronomy Research Collaboratory at the University of Michigan (http://sparc-1.si.umich.edu/sparc/central). This system was designed to facilitate interdisciplinary science, while meeting the needs of disciplinary science and the general public. The Collaboratory consists of a comprehensive and visionary data management system; a set of display, analysis and data mining or information-extraction tools; access to current data; and timely access to older data. These capabilities could facilitate widespread use of seafloor observatory data and data products.

EDUCATION AND PUBLIC OUTREACH

Seafloor observatories will offer an excellent opportunity for educational outreach, as they have the potential to bring the excitement and discovery of ocean science to the public. Virtually all aspects of observatory research are amenable for inclusion in outreach programs, but the most unique are the capabilities for real-time data transmission and, particularly, real-time display of visual images, such as those produced by video, acoustic, and optical sensors.

Outreach efforts can be adapted from successful programs currently in use combined with new approaches designed specifically for observatories. Examples of possible outreach efforts include the following:

- incorporation of real-time image and data feeds into museum exhibits via the Internet;

- incorporation of real-time image and data feeds into K-12 curricula (presented either through schools or museums, as with the Jason program);
- development of curriculum modules for K-12 students that will incorporate research results and scientist profiles;
- establishment of a summer research sabbatical program for K-12 teachers to facilitate interaction with scientists using observatory data;
- development of public World Wide Web sites, publicized through libraries, lay science journals, etc., containing real-time images and data;
- use of public television to broadcast the exciting science coming out of seafloor observatories.

Currently, numerous museum and curriculum outreach programs are being established, but for ocean sciences most of these partnerships are related to coastal studies. An important objective of observatory outreach should be to ensure a broad geographic scale for seafloor observatory public outreach programs.

Involving students and the lay public in the excitement of ocean science needs to be a primary objective of an observatory effort. This effort requires dedicated resources and formal inclusion in project and data management plans. Nationally funded science initiatives that have demonstrated successful outreach programs include NASA, Sea Grant, and IRIS. Other approaches include individual efforts through NSF education programs (the Directorate for Education and Human Resources), local partnerships between individual K-12 schools and nearby marine research institutions or universities, and the privately funded JASON project. Seafloor observatory outreach plans should build on the successes of these programs by incorporating their best and most appropriate aspects. Also, it would be advisable to have a centralized outreach office to coordinate individual efforts and provide information and support. Funding for educational and public outreach efforts should be sought not only from observatory program funds but also from other governmental, private, and industry sources.

6 Findings and Recommendations

In this section, we describe Committee findings, summarize potential benefits and risks associated with the establishment of a seafloor observatory program, and present the Committee's recommendations. Our findings and recommendations are based on input from the January 2000 "Symposium on Seafloor Observatories" held in Islamorada, Florida; various reports and workshop documents made available to the Committee (see Reference List); and the expertise of Committee members.

FINDINGS

Referring to the Statement of Task for this study (Box 1-1), we reach the following overall findings:

1. **Seafloor observatories have significant scientific merit and they will complement and extend current scientific approaches** (see Tables 2-1 through 2-6). It is also recognized, however, that there remain many scientific problems for which this approach is not well suited.

2. **Technical requirements for seafloor observatories.** This report provides the broad technical requirements for a seafloor observatory network based on invited talks, discussions, posters, and working group reports from the "Symposium on Seafloor Observatories." If an observatory infrastructure is put in place, technical requirements will need to be continually refined to address engineering issues that may arise and to allow for enhancement of established capabilities.

3. **Overall feasibility of establishing an infrastructure for seafloor observatories.** Simple mooring and cabled seafloor observatory configurations presently exist, and more complex systems will be feasible in the future if sufficient engineering development resources are devoted to the following major infrastructure elements:

- **Cabled systems**—Depending on the size and complexity of specific networks, significant technical developments are required, especially in the physical design of observatory nodes and power and network management (Chapters 3 and 4).
- **Moored buoys**—Depending on the specific application, significant technical developments are required, especially in satellite telemetry systems, and in the construction of reliable buoy riser systems (Chapters 3 and 4).
- **Ships and Remotely Operated Vehicles (ROVs)**—Off-the-shelf capabilities exist for the installation and maintenance of seafloor observatories, except for very specific applications. Considering the current stress on the ROV fleet, additional capabilities may be required (Chapter 4).
- **Scientific instrumentation**—To fully leverage the capabilities of a seafloor observatory infrastructure, a significant effort to develop new sensors is required in some disciplines, particularly biology and chemistry (Chapter 4).
- **Autonomous Underwater Vehicles (AUVs)**—Significant technical development is required, depending on the tasks envisioned for specific observatory networks (Chapter 4).
- **Data archiving and distribution**—a fully integrated plan for data archiving and distribution is needed; this plan needs to be developed at an early stage in a seafloor observatory program (Chapter 5).

4. **The extent to which seafloor observatories will address future requirements for conducting multidisciplinary research** is very significant, and essential in some fields (Chapter 2).

5. **The level of support for seafloor observatories within ocean and earth science and the broader scientific community** is strong and, although there are indications of support from the broader scientific community, this interest was not quantified in this study. This strong support assumes, however, that seafloor observatories are one part of a broader research strategy, and that adequate support should be provided for a variety of complementary approaches (e.g., traditional ship-based expeditionary research, satellites, drifters, and floats). For example, planetary scientists have expressed an interest in using seafloor observatories as test beds and scientific analogs for exploring oceans on other planetary bodies.

BENEFITS AND RISKS

The establishment of a network of seafloor observatories will represent a new direction in ocean science research, and one that will require a major investment of resources over many decades. Such a major commitment carries with it both potential benefits and risks.

Potential Benefits

The potential benefits associated with the establishment of a seafloor observatory program include the following:

- establishment of a foundation for new discoveries and major advances in the ocean sciences by providing a means to carry out fundamental research on natural and human-induced change on timescales ranging from seconds to decades;
- advances in societally relevant areas of oceanographic research, such as marine biotechnology, the ocean's role in climate change, the evaluation of mineral and fishery resources, and the assessment and mitigation of natural hazards, such as earthquakes, tsunamis, and harmful algal blooms;
- improved access to oceanographic and geophysical data, enabling researchers anywhere in the world to study the oceans and earth in real-time or near real-time by providing basic observatory infrastructure with a wide variety of sensors;
- establishment of permanent observation sites over the 70 percent of Earth's surface covered by oceans to provide truly global geophysical and oceanographic coverage not possible with observations limited to continental or island stations;
- development of new experimental approaches and observational strategies for studying the deep sea;
- enhancement of interdisciplinary research for improving the understanding of interactions between physical, biological, and chemical processes in the oceans;
- establishment of observational resources as fully funded facilities, with the use and access to these facilities being determined by peer-reviewed proposals; and
- increased public awareness of the oceans by providing new educational opportunities for students at all levels using seafloor observatories as a platform for public participation in real-time experiments.

POTENTIAL RISKS

The potential risks associated with the establishment of a seafloor observatory program include the following:

- installation of poorly designed and unreliable observatory systems if program and project planning and risk management are inadequate, technical expertise is lacking, and/or engineering development resources are insufficient;
- potential for interference between experiments resulting from inadequate design, coordination, and/or testing of scientific instrumentation;
- inefficient use of resources if important technological questions are not adequately resolved before major investments in observatory infrastructure are made;
- possible compromise in system performance if critical technologies (e.g., satellite telemetry systems and development of some sensor types) driven by market forces outside the scientific community are not available when needed;
- potential for a growing concentration of technical groups and expertise at a smaller number of institutions involved in supporting the observatories, with the result that many students and scientists may become further removed from understanding how observations are made;
- unreasonable constraints on the freedom of individual investigators to choose the location and timing of their experiments;
- potential for severe impacts on observatory science funding, and funding for other kinds of research and expeditionary science, if the cost of building, maintaining, and operating an observatory infrastructure is higher than initially estimated, and/or if there is a catastrophic loss of observatory components;
- underuse of observatory infrastructure if insufficient funds are budgeted for supporting observatory-related science and the development of scientific instrumentation; and
- potential inability of the present funding structure (based on peer-reviewed, 2- to 5-year duration, discipline-based grants) to judge the merit of projects requiring sustained time-series observations over many years or decades and/or projects that are highly interdisciplinary.

RECOMMENDATIONS

Based on a detailed consideration of the potential benefits and risks that might be associated with a seafloor observatory program, the Committee makes the following recommendations.

1. **The National Science Foundation (NSF) should move forward with the planning and implementation of a seafloor observatory program.**

Observatories represent a promising approach for advancing basic research in the earth and ocean sciences and for addressing societally important issues. This report, which is based on symposium working reports and discussions, documents some of the significant opportunities for new discoveries and major scientific advancements that could result from the establishment of a seafloor observatory network. The establishment of a major seafloor observatory program will require some philosophical and intellectual reorientation within the oceanographic community, building on and complementing the more traditional focus on ship-based mapping and sampling programs. It will also require a major new commitment of resources. Based on the limited information available to the Committee, it is estimated that the initial cost of establishing a seafloor observatory infrastructure could eventually approach several hundred million dollars, and the cost of operating and maintaining this system could be several tens of millions of dollars annually. Thus, seafloor observatories may ultimately require a level of support comparable to that of the present academic research fleet. An investment of this size must be approached cautiously. In addition, mechanisms need to be put in place to deal with contingencies that arise (NRC, 1999) during the establishment of a seafloor observatory network.

2. **A comprehensive seafloor observatory program should include both cabled and moored-buoy systems. Moored-buoy systems should include both relatively high-power, high-bandwidth buoys and simpler, lower-power, limited-bandwidth buoys.**

The diverse applications for seafloor observatory science require the use of both cabled and moored-buoy observatories. Thus, the development of both systems should proceed in parallel. Because of the scientific need to study transient events, it is also important that some rapidly deployable (within weeks or months) observatory systems be developed.

Applications demanding high-telemetry bandwidth and large amounts of power will preferentially use submarine telecommunications cables. Retired telecommunications cables may become available from time to time in areas of scientific interest and may be used on an opportunistic basis. Alternatively, cables may be deployed specifically for scientific purposes as part of a seafloor observatory program. While the high bandwidth and power capabilities of submarine cables make them ideal for seafloor observatories, their relatively high cost could limit the number of such observatories and may restrict their location to areas relatively close to land.

Moored-buoy observatories will be the preferred approach at remote sites where cabled observatories would be prohibitively expensive, when bandwidth and telemetry requirements are modest, or when the duration of the experiment is not sufficiently long to justify the cost of fiber-optic cable. Some applications will require buoy systems with telemetry bandwidths of a few 100s of Mb/day and power generation capabilities of several kW, while in many other cases buoys capable of delivering only a few 10s of W to the seafloor and transmitting a few Mb/day of data to shore could meet scientific needs.

3. **The first step in establishing a seafloor observatory system should be the development of a detailed, comprehensive program and project implementation plan, with review by knowledgeable, independent experts. Program management should strive to incorporate the best features of previous and current large programs in the earth, ocean, and planetary sciences.**

The development of a program and project implementation plan must begin early in the planning process and should include a comprehensive definition of the management and science advisory structure for an observatory program, an implementation timeline and task list with specific milestones, a funding profile for the program, and a schedule for periodic review of program planning and implementation efforts by knowledgeable, independent experts. The management structure must ensure fair and equitable access to observatory infrastructure and must also provide information concerning such issues as protocols and engineering requirements for attaching instruments to a node. The requirements of a management and operational structure for a seafloor observatory program are likely to be similar to other large, coordinated programs in the earth, ocean, and planetary sciences (e.g., University-National Oceanographic Laboratory System [UNOLS], Joint Oceanographic Institutions for Deep Earth Sampling/Ocean Drilling Program [JOIDES/ODP], University Corporation for Atmospheric Research/National Center for Atmospheric Research [UCAR/NCAR], Incorporated Research Institutions for Seismology [IRIS], and National Aeronautic and Space Administration [NASA] mission structures), and the most successful features of these structures should be adopted.

4. **A phased implementation strategy should be developed, with adequate prototyping and testing, before deployment of seafloor observatories on a large scale because of the cost, complexity, and technical challenges associated with the establishment of these systems.**

Great care is needed in the design and implementation of seafloor observatories if they are to meet their scientific potential. Observatory networks should start with simpler nodes having minimal technical risk, adding more complex nodes and configurations over time. This growth plan will allow lessons learned from early deployments to be factored into the design of the later, more complex systems. As such, consideration should be given to the establishment of one or more pilot observatory sites to test prototype systems and sensors in areas that are readily accessible by ships and ROVs. A certified testing capability is also needed to test instrumentation and identify possible interference problems. The engineering development of some component systems that will be part of the initial pilot observatories (such as power and communications systems, AUVs, and sensor technology) could occur in parallel with the establishment of a program and project implementation plan. This will prevent delays in the establishment of initial pilot observatory needs.

5. **A seafloor observatory program should include funding for three essential elements: basic observatory infrastructure, development of new sensor and AUV technology, and scientific research using seafloor observatory data.**

Advances in sensor and AUV development must proceed in parallel with the development, design, manufacture, and installation of basic observatory infrastructure. The development of biological and chemical sensors and instrumentation for long-term in situ measurements is of particularly high priority. AUV development is important because these vehicles will provide the means to greatly expand the footprint of a fixed observatory node by undertaking a variety of mapping and sampling missions around the node. For AUVs to be routinely used at seafloor observatories, significant engineering development is required to provide a reliable docking capability (including homing, capture, data downloading, battery recharging, and mission programming) and the capability to operate for extended periods (at least one year and preferably longer) without human servicing. Some of this development may result in the near-term from the expanding commercial AUV service industry activities in offshore oil and oceanographic research. These developments must be carefully monitored and complemented with appropriate research and development where necessary to meet the planned seafloor observatory mission requirements.

There will be no benefit to seafloor observatories unless scientists are using them to advance our knowledge and understanding of the oceans. Funding for infrastructure and sensors must be balanced with funding for excellent peer-reviewed science that takes advantage of the unique capabilities of observatories. As seafloor observatories are established there will inevitably be some shift in emphasis of existing major science programs and core-supported projects toward more observatory-type studies, but there will also be a need for new support of observatory-related science initiatives. There will still be many important scientific problems that are best addressed using traditional ship-based techniques or fleets of drifters or floats. There is some concern in the scientific community that funding for a seafloor observatory program might have a negative impact on these other ocean-sciences research needs. It is essential that a seafloor observatory program be only one component of a much broader ocean research strategy and that adequate support be provided for a variety of complementary approaches.

6. **New mechanisms should be developed for the evaluation and funding of science proposals requiring sustained time-series observations over many years or decades and for proposals that are highly interdisciplinary.**

Support for observatory-related science will pose new challenges for funding such agencies as NSF. Observatory proposals will typically be more interdisciplinary and will require funding over longer time periods than is currently the norm. The NSF program and review structure is currently not well structured to handle these kinds of proposals, and NSF is taking steps to remedy this problem through increased cooperation and cooperative review of interdisciplinary proposals among program units. This deficiency was highlighted in the National Research Council (NRC) report *Global Ocean Science: Toward an Integrated Approach*, which proposed the creation of a new unit within the Research Section of the NSF Ocean Sciences Division that would be charged with managing a broad spectrum of interdisciplinary projects (NRC, 1999).

7. **A mechanism should be developed to transition successful instrumentation developed by an individual scientist toward a community asset.**

The development of mechanisms to support and encourage the transition of new instrumentation and technology from successful prototype to supported elements of the observatory infrastructure provides a major challenge both for funding agencies and the scientific community. NSF has an excellent track record of funding individual investigators to develop new instrumentation, but turning this instrumentation into a community asset has not been easy. The fundamental problem is to take an instrument that is successful in the hands of the developer and to make it successful in the hands of the broader community. Such a transition may impose significant demands beyond those that an individual investigator would normally undertake; for example, re-engineering elements of the system for non-expert use, production of systems in quantity, and operational support of fielded systems. The importance of operational support can be understood by considering a straightforward Conductivity, Temperature, and Depth (CTD) capability, which requires a trained and experienced operator base despite the commercial availability of

CTD hardware. Involvement of industry through existing mechanisms, such as the Small Business Innovation Research (SBIR) program, although clearly important, does not address the fundamental need of providing mechanisms for extending infrastructure support to appropriate new systems. Since the observatory can be characterized as an extension of our network and power infrastructure into the ocean, with the goal of supporting diverse oceanographic instruments, mechanisms for making such instruments widely available must be an integral part of the plan.

8. **An active public outreach and education program (including K-12) should be a high-priority component of a seafloor observatory program, with a specified percentage of program funding dedicated to this effort.**

Seafloor observatories with real-time communications capabilities should offer an excellent opportunity for public outreach and innovative education initiatives at all levels. With real-time data links to deep sea instruments, it will be possible to involve students and the public directly in ocean science. However, the educational and public outreach potential of seafloor observatories can only be realized by making a meaningful financial commitment to support the development of new ways to present and interpret data for the nonscientific public. To ensure that this effort is made, a percentage of program funding should be provided for outreach and a mechanism should be put into place to encourage researchers to incorporate education and public outreach activities in science proposals. This financial commitment should be drawn from public, private, and industry sources.

9. **A seafloor observatory program should have an open data policy, and resources should be committed to support information centers for archiving observatory data, generating useable data products, and disseminating information to the general public.**

Routine (facility-generated) information products and data should be archived in a central facility and be made available in as near real-time as possible to all investigators and the general public, ideally through the Internet. Data from experimental sensors or individual investigator-initiated experiments should be made publicly available after quality control procedures have been applied, but within 1-2 years of retrieval. A distributed data management system is desirable, and, to the extent possible, existing data management facilities should be used (NRC, 1999).

10. Seafloor observatory programs in the United States should be coordinated with similar international efforts to the extent that progress in the U.S. program is not inhibited. In addition, the Committee recommends that the potential of an integrated, international observatory program be explored.

The goals of a seafloor observatory program in the United States are closely linked to a number of ongoing international initiatives, such as GEOSCOPE and the International Ocean Network. Where practical, coordination of these efforts at the international level will be beneficial and, in some cases, essential. Some major scientific objectives (e.g., global seismic coverage) will be achievable only through this kind of global cooperation.

The vision of establishing a global network of seafloor observatories holds tremendous promise for advancing our understanding of Earth and its oceans. The Committee recognizes that realizing this vision will be difficult and expensive, but based on examination of the important scientific questions that remain to be answered and the current state of technology, the Committee believes the time has come to take the first concrete steps.

References

Baker, P., and M. McNutt, comps. 1996. Report of a Workshop. Ashland Hills, Oregon, December 5-7, 1996. *The Future of Marine Geology and Geophysics (FUMAGES).* Sponsored by the National Science Foundation's Marine Geology and Geophysics and Ocean Drilling Programs, Division of Ocean Sciences.

Bermuda Biological Station for Research, Inc. (BBSR). 2000. "BBSR Home Page." Web page [accessed 28 April 2000]. Available at http://www.bbsr.edu.

Cragg, B. A., R. J. Parkes, J. C. Fry, A. J. Weightman, P. A. Rochelle, J. R. Maxwell, M. Kastner, M. Hovland, M. J. Whiticar, and J. C. Sample. 1995. The impact of fluid and gas venting on bacterial populations and processes in sediments from the Cascadia Margin accretionary system (Sites 888-892) and the geochemical consequences. *Proceedings of the Ocean Drilling Program, Scientific Results* 146(Part 1):399-411.

Delaney, J. R., D. S. Kelley, M. D. Lilley, D. A. Butterfield, J. A. Baross, W. S. D. Wilcock, R. W. Embley, and M. Summit. 1998. The quantum event of oceanic crustal accretion: Impacts of diking at mid-ocean ridges. *Science* 281:222-30.

Dewey, D. B. 1974. Large Navigational Buoy Gen-Set Test Program. *Diesel and Gas Turbine Progress*. November 1974.

Duennebier, F. 2000. Electronic mail message from author on May 9 indicated that this figure will be published in a paper by Caplan-Auerbach.

Ecology and Oceanography of Harmful Algal Blooms (EcoHAB). 1994. "U.S. National Research Plan for the Study of Harmful Algal Blooms." Web page [accessed 28 April 2000]. Available at http://www.redtide.whoi.edu/hab/nationplan/nationplan.html.

Frye, D., K. von der Heydt, M. Johnson, A. Maffei, S. Lerner, and B. Butman. 1999. New Technologies for Coastal Observatories. *Sea Technology* 40(10):29-35.

Global Ocean Observing System (GOOS) Project Office. 1999. "Welcome to the World of GOOS." Web page [accessed 19 April 2000]. Available at http://ioc.unesco.org/goos/.

Hawaii-2 Observatory (H2O). 2000. "Hawaii-2 Observatory (H2O)." Web page [accessed 28 April 2000]. Available at http://www.whoi.edu/science/GG/DSO/H2O/.

Hawaii Ocean Time-series Program (HOT). 2000. "Hawaii Ocean Time-series Program (HOT)." Web page [accessed 28 April 2000]. Available at http://hahana.soest.hawaii.edu/hot/hot.html.

Hawaii Undersea Geo-Observatory (HUGO). 1998. "HUGO: The Hawaii Undersea Geo-Observatory." Web page [accessed 28 April 2000]. Available at http://www.soest.hawaii.edu/HUGO/hugo.html.

Intergovernmental Panel on Climate Change (IPCC). 1995. *The Science of Climate Change: Contribution of Working Group I to the Second Assessment of the Intergovernmental Panel on Climate Change.* J. T. Houghton, L. G. Meira Filho, B. A. Callender, N. Harris, A. Kattenberg, and K. Maskell, eds. Cambridge, England: Cambridge University Press.

Jumars, P., and M. Hay, comps. 1999. Report of the OEUVRE Workshop. Keystone, Colorado, March 1-6, 1998. *Ocean Ecology: Understanding and Vision for Research.* Under sponsorship of an award to the University Corporation for Atmospheric Research, Joint Office for Science Support, from the National Science Foundation, Division of Ocean Sciences.

Karl, D. M., and Lukas R. 1996. The Hawaii Ocean Time-series (HOT) Program: Background, Rationale, and Field Implementation. *Deep-Sea Research II.* 43:129-56.

Lukas R., and D. M. Karl. 1998. Hawaii Ocean Time-series (HOT): A Decade of Interdisciplinary Oceanography. *SOEST Technical Report 99-05.* CD-ROM ed. Honolulu, Hawaii: School of Ocean and Earth Science and Technology.

Maritime Communications Services, Inc. (MCS). 2000. "OceanNet." Web page [accessed 8 May 2000]. Available at http://www.mcs.harris.com/oceannet/oceannet.html.

Mayer, L., and E. Druffel, eds. 1999. Report of the FOCUS Workshop. Seabrook Island, South Carolina, January 6-9, 1998. *The Future of Ocean Chemistry in the U.S.* Under sponsorship of an award to the University Corporation for Atmospheric Research, Joint Office for Science Support, from the National Science Foundation, Division of Ocean Sciences.

Montagner, J. P., and Y. Lancelot, eds. 1995. Multidisciplinary observatories on the deep seafloor. Conference Report. INSU/CNRS, IFREMER, ODP-France, OSN/USSAC, ODP-Japan.

Moore, G. and C. Moore, eds. 1998. "The Seismogenic Zone Experiment (SEIZE) Workshop. Waikoloa, Hawaii, June 3-6, 1997. Final Report." Web page [accessed 5 May 2000]. Available at http://www.soest.hawaii.edu/moore/seize/.

National Research Council (NRC). 1992. *Oceanography in the Next Decade: Building New Partnerships.* Washington, D.C.: National Academy Press.

———. 1994a. *GOALS (Global Ocean-Atmosphere-Land System) for Predicting Seasonal-to-Interannual Climate: A Program of Observation, Modeling, and Analysis*. Washington, D.C.: National Academy Press.

———. 1994b. *Review of U.S. Planning for the Global Ocean Observing System*. Washington, D.C.: National Academy Press.

———. 1996a. *Natural Climate Variability on Decade-to-Century Time Scales*. Washington, D.C.: National Academy Press.

———. 1996b. *Undersea Vehicles and National Needs*. Washington, D.C.: National Academy Press.

———. 1997. *The Global Ocean Observing System: Users, Benefits, and Priorities*. Washington, D.C.: National Academy Press.

———. 1998a. *Opportunities in Ocean Sciences: Challenges on the Horizon*. Washington, D.C.: National Academy Press.

———. 1998b. *Decade-to-Century-Scale Climate Variability and Change: A Science Strategy*. Washington, D.C.: National Academy Press.

———. 1998c. *A Scientific Strategy for U.S. Participation in the GOALS (Global Ocean-Atmosphere-Land System) Component of the CLIVAR (Climate Variability and Predictability) Programme*. Washington, D.C.: National Academy Press.

———. 1999. *Global Ocean Science: Toward an Integrated Approach*. Washington, D.C.: National Academy Press.

———. 2000. *Bridging Boundaries Through Regional Marine Research*. Washington, D.C.: National Academy Press.

North East Pacific Time-series Undersea Networked Experiments (NEPTUNE). 2000. "NEPTUNE: A Fiber-Optic Telescope to Inner Space." Web page [accessed 19 April 2000]. Available at http://www.neptune.washington.edu/.

Nowlin, W.D., Jr. 1999. A strategy for long-term observations. *Bulletin of the American Meteorological Society* 80:621-27.

Ocean Seismic Network (OSN), sponsor. 1995. A Workshop Held at Scripps Institution of Oceanography, La Jolla, California, on February 14-15, 1995. *Broadband Seismology in the Oceans: Towards a Five-Year Plan*. Washington, D.C.: Joint Oceanographic Institutions, Inc.

OCEANOBS99, SOC Secretariat. 1999. "OCEANOBS99 Program." Web page, [accessed 14 March 2000]. Available at http://www.bom.gov.au/OceanObs99/Program.html.

Pacific Marine Environmental Laboratory (PMEL). 1999. "ATLAS Buoy Information." Web page [accessed 4 May 2000]. Available at http://www.pmel.noaa.gov/toga-tao/buoy.html.

Purdy, G. M., and A. M. Dziewonski, convenors. 1988. Proceedings of a Workshop on Broadband Downhole Seismometers in the Deep Ocean. Woods Hole Oceanographic Institution, Woods Hole, Massachusetts, April 26-28, 1988. Sponsored by the Joint Oceanographic Institutions, Inc./U.S. Science Advisory Committee.

Royer, T., and W. Young, comps. 1999. Report of the APROPOS Workshop. Monterey, California, December 15-17, 1997. *The Future of Physical Oceanography*. Under sponsorship of an award to the University Corporation for Atmospheric Research, Joint Office for Science Support from the National Science Foundation, Division of Ocean Sciences.

Rutgers University-Coastal Ocean Observation Lab (COOL). 1999. "Rutgers University Long-Term Ecosystem Observatory." Web page [accessed 10 May 2000]. Available at http://marine.rutgers.edu/mrs/coolresults/1999/tos2/figure7.jpg.

Summerhayes, C. P. 1996. Ocean Resources. In *Oceanography: An Illustrated Guide*. C. P. Summerhayes, and S. A. Thorpe, eds. London: Manson Publishing Ltd. pp. 314-37.

Summit, M., and J. A. Baross. 1998. Thermophilic subseafloor microorganisms from the 1996 North Gorda Ridge eruption. *Deep-Sea Research II*. 45:2751-66.

University of Miami, Rosenstiel School of Marine and Atmospheric Science. 1999. "DEOS Home Page." Web page [accessed 3 May 2000]. Available at http://www.geodesy.miami.edu/deos.html.

ADDITIONAL REFERENCES

(Not Cited in Text)

Buddenberg, R. A., A. C. Chave, H. M. Frazier, Jr., B. Howe, H. Kirkham, A. R. Maffei, R. A. Petitt Jr., D. H. Rodgers, F. Beecher Wooding, and D. R. Yoerger (NEPTUNE Steering Committee). 2000. *NEPTUNE Technical Summary*. Material extracted from the NEPTUNE Feasibility Study.

Butler, R. and T. E. Pyle, convenors. 1990. *Workshop on Scientific Uses of Undersea Cables. Honolulu, Hawaii, January 30 - February 1, 1990*. A. D. Chave, R. Butler, and T. E. Pyle, eds. Washington, D.C.: Joint Oceanographic Institutions, Inc. Report to the National Science Foundation, National Oceanic and Atmospheric Administration, Office of Naval Research, and the U.S. Geological Survey.

Delaney, J., A. Chave, R. Heath, and B. Howe. 2000. Dynamics of Earth and Ocean Systems. *Neptune: An Interim Report on the Plate-Scale Observatory Concept*. Supported by the National Oceanographic Partnership Program (NOPP) and the NEPTUNE partners (University of Washington, Woods Hole Oceanographic Institution, Pacific Marine Environmental Laboratory, and Jet Propulsion Laboratory). Prepared for the National Research Council's Ocean Studies Board Symposium on Seafloor Observatories: Challenges and Opportunities, Islamorada, Florida, January 10-12, 2000.

Incorporated Research Institutions for Seismology (IRIS). 1994. *Science Plan for A New Global Seismographic Network*. Washington, D.C.: IRIS, Incorporated.

Incorporated Research Institutions for Seismology (IRIS), United States Geological Survey (USGS), and National Science Foundation (NSF). 1990. *Technical Plan for a New Global Seismographic Network*. Washington, D.C.

Joint Oceanographic Institutions (JOI) ad hoc Committee. 1994. *Dual Use of IUSS: Telescopes in the Ocean*. Washington, D.C.: Joint Oceanographic Institutions, Inc.

Kasahara, J., and A. Chave, convenors (Committee for Scientific Use of Submarine Cables, Japan and the U.S. Steering Committee for Scientific Use of Undersea Cables). 1997. *Marine Geophysical Research Using Undersea Cables*. H. Utada, K. Noguchi, C. Harayama, and N. Natsushima, eds. Proceedings of International Workshop on Scientific Use of Submarine Cables. Okinawa, Japan, February 25-28, 1997. Okinawa, Japan: Japan Print Center.

National Oceanographic Partnership Program (NOPP). 1999. *Toward a U.S. Plan for an Integrated, Sustained Ocean Observing System*. A report prepared on behalf of the National Ocean Research Leadership Council (NORLC).

Orcutt, J. and A. Schultz, Working Group Co-Chairs. 1999. *DEOS Global Working Group Report: Moored Buoy Ocean Observatories*.

Wells, N. C., W. J. Gould, and A. E. S. Kemp. 1996. Chapter 3: The Role of Ocean Circulation in the Changing Climate. *Oceanography: An Illustrated Guide*. C. P. Summerhayes and S. A. Thorpe. London, U.K.: Manson Publishing Ltd., pp. 41-58.

Appendixes

Steering Committee and Staff Biographies

Chairman

William Ryan received his Ph.D. from Columbia University in 1971. He is currently a Senior Scientist at the Lamont-Doherty Earth Observatory and an Adjunct Professor of Earth and Environmental Sciences at Columbia University. Dr. Ryan is an expert in the use of multi-frequency sonar systems for seabed mapping and seafloor characterization, while also being active in a wide range of other earth science interests.

Vice Chairman

Robert Detrick obtained his Ph.D. in 1978 from the Massachusetts Institute of Technology/Woods Hole Oceanographic Institution joint program in Marine Geophysics. Dr. Detrick has worked as a Senior Scientist at the Woods Hole Oceanographic Institution (WHOI) since 1991. His research focuses on the seismic structure and evolution of oceanic crust, mid-ocean ridge processes, and the dynamics of the oceanic upper mantle. Dr. Detrick is currently the Chair of the Department of Geology and Geophysics at Woods Hole.

Keir Becker received his Ph.D. in oceanography from the Scripps Institution of Oceanography in 1981, and he is currently a Professor in the Division of Marine Geology and Geophysics at the University of Miami, Rosenstiel School of Marine and Atmospheric Science. Dr. Becker's research interests include the study of heat flow and hydrothermal circulation through the ocean crust and how this flow depends on and alters crustal physical properties. Dr. Becker has extensive experience in observatory research, particularly in the use of borehole observatories.

James Bellingham received his Ph.D. in physics from the Massachusetts Institute of Technology in 1988. Dr. Bellingham is currently Director of the Engineering Division of the Monterey Bay Aquarium Research Institute and, formerly, the Laboratory Manager for the Autonomous Underwater Vehicles (AUV) Laboratory of the MIT Sea Grant College Program. Dr. Bellingham is a highly respected ocean engineer with extensive experience with Autonomous Undersea Vehicles (AUVs). Dr. Bellingham has been a member of the University-National Oceanographic Laboratory System (UNOLS) Deep Submergence Science Committee since 1993.

Roger Lukas obtained his Ph.D. in oceanography from the University of Hawaii in 1981 and is currently a Professor at the University of Hawaii, School of Ocean and Earth Science and Technology. His research focuses on ocean-atmosphere interaction, seasonal-to-interannual climate variability, tropical ocean currents, equatorially trapped waves, and the distribution of oceanic water mass properties in the tropics and subtropics. Dr. Lukas is currently a member of the Ocean Studies Board.

John Lupton obtained his Ph.D. in physics from the California Institute of Technology in 1971. Dr. Lupton is an Oceanographer with the NOAA Pacific Marine Environmental Laboratory and is a member of the NOAA Hydrothermal Vents System (VENTS) research group. His research addresses gas and fluid chemistry of submarine hydrothermal systems and the application of ocean tracer distributions to ocean circulation studies.

Lauren Mullineaux obtained her Ph.D. in oceanography from the Scripps Institution of Oceanography in 1987. Dr. Mullineaux is both the Biology Department Education Coordinator and an Associate Scientist at the Woods Hole Oceanographic Institution. Her research focuses on benthic community ecology and the colonization and dispersal of deep-sea hydrothermal vent species. Dr. Mullineaux is presently a member of the InterRidge Steering Committee and chair of their biology subcommittee.

Jack Sipress received his doctorate in electrical engineering from Polytechnic University, Brooklyn, in 1961. Dr. Sipress is currently the President of Sipress Associates. His specialty is the development and implementation of international communications facilities via undersea lightwave cables. Over a 20-year period, he provided technical leadership for the planning, research, design, development, manufacture, installation, and maintenance of undersea systems and associated technologies at Bell Laboratories, AT&T Submarine Systems Inc., and Tyco Submarine Systems Ltd. Dr. Sipress is a member of the National Academy of Engineering.

NATIONAL RESEARCH COUNCIL STAFF

Alexandra Isern is a Program Officer with the Ocean Studies Board. She received her Ph.D. in Marine Geology from the Swiss Federal Institute of Technology in 1993. Dr. Isern was a lecturer in Oceanography and Geology at the University of Sydney, Australia, from 1994-1999. Her research focuses on the influences of paleoclimate and seal-level variability on the development of carbonate platforms. Dr. Isern is co-chief scientist for Ocean Drilling Program Leg 194, which will investigate the magnitudes of ancient sea-level change.

Shari Maguire serves as a Research Assistant with the Ocean Studies Board. She received her B.A. from Miami University in International Studies and Russian in 1994. Currently, Ms. Maguire is studying the biological sciences at the University of Maryland and plans to pursue a graduate degree in the health care field.

Megan Kelly is a Project Assistant with the Ocean Studies Board of the National Research Council. She received her B.S. in Marine Science from the University of South Carolina in May 1999 and plans to pursue a graduate degree in marine policy.

Symposium Participants

Tim Ahern	*IRIS*
Stace Beaulieu	*Woods Hole Oceanographic Institution*
Keir Becker	*University of Miami*
Robin Bell	*Lamont-Doherty Earth Observatory*
James Bellingham	*Monterey Bay Aquarium Research Institution*
Kenneth Brink	*Woods Hole Oceanographic Institution*
Melbourne Briscoe	*Office of Naval Research*
Dave Butterfield	*NOAA-Pacific Marine Environmental Laboratory*
Alan Chave	*Woods Hole Oceanographic Institution*
Dayle Chayes	*Lamont-Doherty Earth Observatory*
Larry Clark	*National Science Foundation*
Kim Cobb	*Scripps Institution of Oceanography*
Robert Collier	*Oregon State University*
James Cowen	*University of Hawaii*
John Delaney	*University of Washington*
Robert Detrick	*Woods Hole Oceanographic Institution*
Tommy Dickey	*University of California, Santa Barbara*
Tim Dixon	*University of Miami*
Fred Duennebier	*University of Hawaii*
Adam Dziewonski	*Harvard University*
Robert Embley	*NOAA-Pacific Marine Environmental Laboratory*
Charles Eriksen	*University of Washington*
Chris Fox	*NOAA-Pacific Marine Environmental Laboratory*
Rana Fine	*University of Miami*
Charles Fisher	*Pennsylvania State University*
Daniel Frye	*Woods Hole Oceanographic Institution*

Chris German	*Southampton Oceanography Centre*
Morgan Gopnik	*National Research Council*
Stephen Hammond	*NOAA-Pacific Marine Environmental Laboratory*
Ross Heath	*University of Washington*
John Hildebrand	*Scripps Institution of Oceanography*
Bruce Howe	*University of Washington*
Alexandra Isern	*National Research Council*
William Johns	*University of Miami*
Kenneth Johnson	*Monterey Bay Aquarium Research Institution*
Terrence Joyce	*Woods Hole Oceanographic Institution*
Peter Jumars	*University of Maine*
Miriam Kastner	*Scripps Institution of Oceanography*
Megan Kelly	*National Research Council*
Michael Kelly	*National Oceanic and Atmospheric Administration*
Gary King	*University of Maine*
Gary Klinkhammer	*Oregon State University*
Herb Kroehl	*NOAA-National Geophysical Data Center*
Larry Langebrake	*University of South Florida*
Marvin Lilley	*University of Washington*
Roger Lukas	*University of Hawaii*
John Lupton	*NOAA-Pacific Marine Environmental Laboratory*
Douglas Luther	*University of Hawaii*
Marcia McNutt	*Monterey Bay Aquarium Research Institution*
Anthony Michaels	*University of Southern California*
Peter Mikhalevsky	*Science Applications International Corporation*
Steven Miller	*National Undersea Research Center*
Sujata Millick	*Office of Naval Research*
Lauren Mullineaux	*Woods Hole Oceanographic Institution*
John Orcutt	*Scripps Institution of Oceanography*
Kurt Polzin	*Woods Hole Oceanographic Institution*
Michael Purdy	*National Science Foundation*
Keith Raybould	*Monterey Bay Aquarium Research Institution*
Anna-Louise Reysenbach	*Portland State University*
Barbara Romanowicz	*University of California, Berkeley*
William B. F. Ryan	*Lamont-Doherty Earth Observatory*
Henrik Schmidt	*Massachusetts Institute of Technology*
Oscar Schofield	*Rutgers University*
Adam Schultz	*Southampton Oceanography Centre*
Jack Sipress	*Sipress Associates*
Uwe Send	*University of Kiel*
Ken Smith	*Scripps Institution of Oceanography*
Ralph Stevens	*Woods Hole Oceanographic Institution*
Kiyoshi Suyehiro	*Japan Marine Science and Technology Center*

Meg Tivey	*Woods Hole Oceanographic Institution*
Patrick R. Trischitta	*Tyco Submarine Systems*
Martin Visbeck	*Lamont-Doherty Earth Observatory*
Karen von Damm	*University of New Hampshire*
Robert Weller	*Woods Hole Oceanographic Institution*
Will Wilcock	*University of Washington*
Roy Wilkens	*Office of Naval Research*
Carl Wunsch	*Massachusetts Institute of Technology*
Dana Yoerger	*Woods Hole Oceanographic Institution*

Symposium Program

SUNDAY, JANUARY 9

5:00 p.m. Registration

5:30 p.m. Ice-breaker and posters related to seafloor observatories

MONDAY, JANUARY 10

7:30 a.m. Breakfast and Registration

OPEN SESSION

8:30 a.m. Introductory remarks – William Ryan, Chair, NRC Steering Committee

9:00 a.m. Long-term Eulerian measurements for the ocean circulation
Carl Wunsch (MIT)

9:20 a.m. A vision for future research of oceanic ecosystems and carbon cycling using seafloor observatories
Stace Beaulieu (WHOI)

9:40 a.m. Insights into seafloor hydrothermal systems from monitoring responses to natural perturbations
Meg Tivey (WHOI)

10:00 a.m. Interdisciplinary ramifications of seafloor microbial observatories
Anna-Louise Reysenbach (Portland State University)

10:20 a.m. Break

10:40 a.m. Coastal oceanography and time-series measurements
Peter Jumars (University of Maine)

11:00 a.m. The need for an ocean floor network in global earth studies
Adam Dziewonski (Harvard)

11:20 a.m. Introduction to breakout session 1

11:40 a.m. Breakout session 1 and working lunch

3:30 p.m. Break

3:45 p.m. Breakout session 1 wrap-up (in plenary)

5:15 p.m. Meeting adjourns for the day

5:30 p.m. Cash-bar and posters related to seafloor observatories

7:00 p.m. Symposium Dinner

8:30 p.m. Evening speaker: Neptune: Oceanography at the scale of a tectonic plate
John Delaney (University of Washington)

Tuesday, January 11

7: 30 a.m. Breakfast

OPEN SESSION

8:30 a.m. Summary of previous day and introduction to the day's activities – William Ryan

8:40 a.m. Undersea fiberoptic cable systems: High-tech telecommunications tempered by a century of ocean cable experience
Patrick Trischitta (Tyco Submarine Systems)

9:00 a.m. A moored-buoy observatory for the deep ocean
Dan Frye (WHOI)

9:20 a.m. Science interfaces for seafloor observatories
Alan Chave (WHOI)

9:40 a.m. Sensor systems for ocean observatories: How do we monitor processes that transform mantle energy into ecosystems?
Ken Johnson (MBARI)

10:00 a.m. Future applications of remotely operated and autonomous vehicles in deep sea observatories
Dana Yoerger (WHOI)

10:20 a.m. Break

10:40 a.m. Successfully managing larger scientific databases faster, cheaper, better?
Herb Kroehl (National Geophysical Data Center)

11:00 a.m. Introduction to breakout session 2 (in plenary)

11:20 a.m. Breakout session 2 and working lunch

2:00 p.m. Breakout session 2 wrap-up; introduction to breakout session 3 (in plenary)

3:00 p.m. Break

3:15 p.m. Breakout session 3

5:30 p.m. Adjourn for the day

WEDNESDAY, JANUARY 12

7:30 a.m. Breakfast

OPEN SESSION

8:30 a.m. Summary of previous day and introduction to the day's activities – William Ryan

8:45 a.m.	Breakout group 3 continued
11:00 a.m.	Breakout session 3 wrap-up (in plenary)
12:30 p.m.	Chairman's summary and closing remarks
1:00 p.m.	Symposium ends
1:00 p.m.	Lunch
2:00 p.m.	Steering committee reconvenes with session leaders to draft session outcomes

Closed Session

4:30 p.m.	Steering committee reconvenes to discuss recommendations

Charges to Breakout Groups

BREAKOUT CHARGES

Themes for Breakout Groups 1 and 2

Group A: Role of the ocean in climate (session chair: Robert Weller)
Group B: Fluids, chemistry, and life in the oceanic crust (session chair: Miriam Kastner)
Group C: Coastal ocean perturbation and processes (session chair: Robin Bell)
Group D: Dynamics of oceanic lithosphere and imaging the earth's interior (session chair: John Orcutt)
Group E: Turbulent mixing and biophysical interaction (session chair: Tommy Dickey)
Group F: Biodiversity and ecosystem dynamics (session chair: Ken Smith)

Breakout Group 1

Charge for Breakout Group 1: What is the potential for seafloor observatories to lead to significant scientific advances?

The goal of this session is to assess the extent to which seafloor observatories will address important scientific issues in earth and ocean sciences and to articulate the scientific merit of establishing observatories on the seafloor. As initial guidance for this discussion, a list of the key scientific problems in ocean sciences has been assembled from the recent NSF "Futures" reports on biological (OEUVRE), chemical (FOCUS), geological (FUMAGES), and physical oceanography (APROPOS). Be visionary! — feel free to amend this list or synthesize where questions appear closely related.

1. For those important scientific questions that are pertinent to your theme, assess the potential for seafloor observatories to make significant scientific contributions. Please use the following rating for seafloor observatories as potential tools to address the scientific issues:

E = Essential (without seafloor observatories, breakthrough advances on this issue cannot be made)
U = Useful (seafloor observatories will provide information that is useful and important, but not sufficient to resolve important issues)
N = Not needed or not appropriate

In rating the scientific potential for seafloor observatories, please amplify with comments to justify your results.

Please list any scientific questions that arose from your discussions which were not on the lists below.

2. For the scientific questions where observatories were felt to be essential or useful, what are some specific advances or breakthroughs that you would anticipate to result from seafloor observatory capabilities?

3. What are the strategic impacts on research methodology that will result from establishment of seafloor observatories?

BREAKOUT GROUP 2

Charge for Breakout Group 2: Which observatory type is best suited to investigate the scientific problems identified in Breakout Group 1?

The aim of this breakout session is to begin discussing the technical needs for seafloor observatories within the context of the scientific problems deemed to benefit from observatory science.

1. For each of the scientific problems that were classified as benefiting from observatory science in Breakout Group 1, identify which of the observatory types listed below would be best suited to answer the scientific question. Please justify your answer.

Relocatable Observatory: An observatory that is expected to be installed at a site for a limited period of time (up to a few years) and capable of then being redeployed elsewhere. Although cable connectivity to shore may be attractive for some applications, in all probability this class of observatory will be mooring based with satellite or "rf" communications to shore installations and a communication/power riser from the seafloor to the ocean surface. The relocatable observatory may support an array of devices on the seafloor that are acoustically, electrically, or fiber-optically linked back to the mooring, as well as AUVs and their docking stations.

Long-term Observatory: An observatory which is expected to be installed at a site for a period of decades or more. Although a mooring-based installation may be attractive for some applications, in all probability this class of observatory will utilize an undersea cable from the shore to provide power and communications to one or more "nodes" and will support science experiments with power and communications needs not economically or logistically supportable by a mooring-based observatory. An individual node might support a range of devices, such as remotely deployed seafloor systems and AUV docking stations. A permanent observatory might have a large number of nodes.

Global/Basin-Scale Observatory Network: An observatory designed to provide basin or global-scale coverage through a network of observatory nodes. The individual nodes might be mooring or cable-based. Many more advanced options might also be a part of such an observatory. These include the incorporation of long-endurance mobile vehicles to augment the Eulerian network with Lagrangian and additional Eulerian observations at locations at a distance from the main node.

2. While completing task 1 above, keep in mind the following questions and comment on those that you feel are appropriate.

- What would be the distribution of sites required to address scientific problems within the overall theme assigned to your group?
- What would be the optimal frequency for data collection in order to characterize the events being investigated?
- What would be the scientific approach if the event in question turns out to be non-periodic? What strategies can be employed to best guarantee that this event can be studied?
- What types of data would be needed to solve the scientific problems being discussed (regardless or whether these measurements are currently possible or not)?
- Would the scientific problems you are addressing require real-time data telemetry?
- What sensors are currently available to collect the above-mentioned data sets? What kinds of sensors would be beneficial to develop?

BREAKOUT GROUP 3

Relocatable Observatory

session leaders: Keith Raybould and Chris Fox

An observatory that is expected to be installed at a site for a limited period of time (up to a few years) and capable of then being redeployed elsewhere. Although cable connectivity to shore may be attractive for some applications, in all probability this class of observatory will be mooring-based with satellite or rf communications to shore installations and a communication/power riser from the seafloor to the ocean surface. The relocatable observatory may support an array of devices on the seafloor that are acoustically, electrically, or fiber-optically linked back to the mooring, as well as AUVs and their docking stations.

Long-term Observatory

session leaders: Fred Duennebier and Marv Lilley

An observatory that is expected to be installed at a site for a period of decades or more. Although a mooring-based installation may be attractive for some applications, in all probability this class of observatory will utilize an undersea cable from the shore to provide power and communications to one or more "nodes" and will support science experiments with power and communications needs not economically or logistically supportable by a mooring-based observatory. An individual node might support a range of devices, such as remotely deployed seafloor systems and AUV docking stations. A permanent observatory might have a large number of nodes.

Global/Basin-Scale Observatory Network

session leaders: Barbara Romanowicz and Doug Luther

An observatory designed to provide basin or global-scale coverage through a network of observatory nodes. The individual nodes might be mooring or cable-based. Many more advanced options might also be a part of such an observatory. These include the incorporation of long endurance mobile vehicles to augment the Eulerian network with Lagrangian and additional Eulerian observations at locations at a distance from the main node.

Charge for Breakout Group 3: What are the technical requirements needed to establish a series of seafloor observatories?

The aim of this session is to build on the results of Breakout Session 2 and determine specific technical requirements necessary for establishing a series of seafloor observatories to solve the scientific problems discussed in Breakout Group 1.

1. Technical requirements: For each of the scientific themes that would benefit from observatory science discussed in Breakout Group 1, list specific requirements needed to address the following feasibility issues:

- What are the primary technical capability requirements needed (including sensors, power, bandwith, communication)?
- What resources will be needed for deployment of the observatories (e.g., ship-days and ROVs)?
- What are the primary maintenance requirements (including contingencies)?
- What are the primary operational requirements related to the seafloor observatory in question (e.g., technician capabilities and numbers)?
- What technology enhancements and improvements will be required?

While completing this task, keep in mind the following questions and comment on those you feel are appropriate.

Program/Information management

- What would be the most effective management structure for a large seafloor observatory program?

- Seafloor observatories will produce large amounts of raw and processed data. How should these data be managed and made available?

Educational outreach

- What would be some educational "experiments" that could be developed using observatory data within the overall theme you are discussing?

- Often, educational components of large scientific programs are not implemented to their full potential. What structure could be put in place to ensure that educational outcomes are realized within a seafloor observatory program?

2. Discuss potential scientific synergies where multiple scientific problems could be addressed with common infrastructure. An important outcome of Breakout Group 3 will be a synthesis of common requirements that run across themes.

Acronyms and Abbreviations

3-D	Three-Dimensional
AOSN	Autonomous Ocean Sampling Networks
APROPOS	Advances and Primary Research Opportunities in Physical Oceanography Studies
ATLAS	Autonomous Temperature Line Acquisition System
ATOC	Acoustic Thermometry of Ocean Climate
AUV	Autonomous Underwater Vehicle
BATS	Bermuda Atlantic Time-series Study
BOREHOLE	BOREHole Observatories, Laboratories, and Experiments
BTM	Bermuda Testbed Mooring
CLIVAR	CLImate VARiability and predictability programme
CMB	Core-Mantle Boundary
CO_2	Carbon Dioxide
CORK	Circulation Obviation Retrofit Kits
CTD	Conductivity, Temperature, and Depth
DC	Direct Current
DEOS	Deep Earth Observatories on the Seafloor (later changed to Dynamics of Earth and Ocean Systems)
ENSO	El Niño/Southern Oscillation
FUMAGES	Future of Marine Geosciences
GEO	Global Eulerian Observations
GOOS	Global Ocean Observing System
GPS	Global Positioning System
H2O	Hawaii-2 Observatory
HOT	Hawaii Ocean Time-series
HUGO	Hawaii Undersea Geo-Observatory

IMET	Improved METeorological
ION	International Ocean Network
IRIS	Incorporated Research Institutions for Seismology
JGOFS	Joint Global Ocean Flux Study
JOIDES	Joint Oceanographic Institutions for Deep Earth Sampling
LEO-15	Long-term Ecosystem underwater Observatory, 15 meters below the surface
LNB	Large Navigational Buoys
LTER	Long-Term Ecological Research Network
MBARI	Monterey Bay Aquarium Research Institute
NASA	National Aeronautic and Space Administration
NCAR	National Center for Atmospheric Research
NEMO	NEw Millennium Observatory
NEPTUNE	North East Pacific Time-series Undersea Networked Experiments
NODC	National Oceanographic Data Center
NRC	National Research Council
NSF	National Science Foundation
ODP	Ocean Drilling Program
OEUVRE	Ocean Ecology: Understanding and Vision for REsearch
ONR	Office of Naval Research
OOPC	Ocean Observing Panel for Climate
OOSDP	Ocean Observing System Development Panel
OSB	Ocean Studies Board
OSN	Ocean Seismic Network
RIDGE	Ridge InterDisciplinary Global Experiments
ROV	Remotely Operated Vehicle
SBIR	Small Business Innovation Research
SEIZE	SEIsmogenic Zone Experiment (MARGINS, JOIDES)
SOEST	School of Ocean and Earth Science and Technology (University of Hawaii)
SOSUS	SOund SUrveillance System
TAO	Tropical Atmosphere-Ocean array
UCAR	University Corporation for Atmospheric Research
UNOLS	University-National Oceanographic Laboratory System
WHOI	Woods Hole Oceanographic Institution
WOCE	World Ocean Circulation Experiment